C0-ATU-744

DEADLY
Beautiful

DEADLY *Beautiful*

VANISHING KILLERS OF THE ANIMAL KINGDOM

LIANA JOY CHRISTENSEN

ILLUSTRATIONS BY IAN FAULKNER

First published 2011

Exisle Publishing Limited
'Moonrising', Narone Creek Road, Wollombi, NSW 2325, Australia
P.O. Box 60–490, Titirangi, Auckland 0642, New Zealand
www.exislepublishing.com

National Library of Australia Cataloguing-in-Publication Data:

Christensen, Liana Joy

Deadly beautiful : vanishing killers of the animal kingdom /
Liana Joy Christensen.

ISBN 9781921497223 (hbk.)

Includes bibliographical references and index.

Dangerous animals.
Endangered species.

Faulkner, Ian.

574.6

Designed by saso content & design pty ltd
Typeset in Adobe Caslon 12/16.5 by 1000 Monkeys Typesetting Services
Printed in Singapore by KHL Printing Co Pte Ltd

This book uses paper sourced under ISO 14001 guidelines from well-managed forests and other controlled sources.

10 9 8 7 6 5 4 3 2 1

In memory of Val Plumwood. With strong body, fierce intellect and deep integrity you embraced the natural world on its own terms.
May native grasses flourish where you lie.

Contents

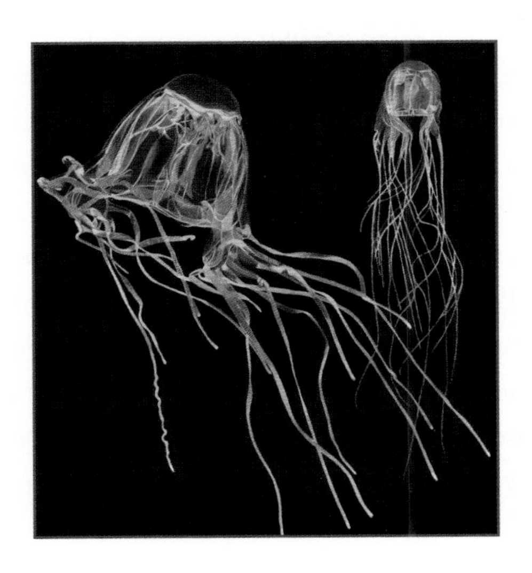

INTRODUCTION

PERSONALLY, I ALWAYS THOUGHT IT WAS FAR MORE SENSIBLE TO fear bees than sharks. Then I met the beekeeper's wife who told me that many more people die from strawberry allergies than bee stings. I've yet to meet a person who has a phobia about strawberries. But countless people have phobias about spiders and snakes, and such phobias are by no means restricted to highly strung non-scientists. Several prominent zoologists in the United

States who specialise in snakes — and freely admit to loving them — count themselves as arachnophobes. This allows them to be sympathetic to ophidiophobes (snake phobics). There are plenty of other animal candidates when it comes to phobias. Still, many humans take a perverse pride in wryly asserting that we are the most dangerous animal of all. As usual, we have inflated our importance in the scheme of things. We are indeed dangerous both to ourselves and to the other animals we share the planet with. But in terms of being the most dangerous, we are humbled by microbes. Billions more people and other animals have been killed by microbial animals than large predator species.

Clearly, then, those animals that act as vectors for viral and bacterial disease pose the greatest threat of all. That is why malarial mosquitoes are ranked as the number one most dangerous animal in the world. Of 400 species of Anopheles mosquitoes, only 30 are considered serious vectors for malaria, but worldwide there are 300 million cases of malaria each year — and 100 million deaths. Generally, though, when people think of deadly and dangerous animals, certain specific groups of larger animals are what spring to mind: bears, sharks, big cats, crocodiles, alligators, wolves are among the usual suspects. The list also includes stingrays, snakes, spiders, scorpions, box jellyfish, stonefish, cone shells and the blue-ringed octopus. These animals are the focus of *Deadly Beautiful.* They are collected together not because of any zoological similarity — they comprise mammals, reptiles, fish, arachnids — but because they can and have killed humans. And because each of them carries the reputation of being a fearsome and deadly animal.

WHAT ARE THE ODDS?

Even the most cursory investigation, however, reveals that the reputation of many deadly and dangerous animals far exceeds the actual danger they pose to humans. Here we enter the strange world of statistics. Sharks, for example, kill an average of 50 or 60 people a year worldwide. To put that in perspective, in the early 21st century around 900 people are murdered each year in New York State alone, and that's down from regular annual highs of around 2500 in the late 1980s to the early 1990s. On the other hand, death — like pregnancy — is an absolute condition. It is hardly a comfort that fewer than one in 100 million people is bitten by a funnel-web spider if you are the relative of somebody who was. Nonetheless, realistic risk assessment is important. Even the most puritanical of us are gamblers — we play the odds every time we get into a car.

Whether you are fearful of — or fascinated by — deadly animals, this book will give you the chance to find out more about this special class of creatures and the way humans deal with them. It will soon become evident that paying attention to dangerous and deadly animals tells us as much about human nature as it does about the animals themselves. Snakes, spiders, scorpions, bears, crocodiles, wolves, lions and tigers populate our mythologies and fables. Sometimes good, sometimes evil, always powerful: these über-animals have a second life in human imagination. They also have a real life on planet Earth. Our responses to the actual animals and the real and exaggerated threats they pose are as wildly varied as the creatures themselves.

USUAL AND UNUSUAL SUSPECTS

Zoologists and amateur naturalists generally have a realistic regard for the dangers associated with the subjects of their studies, together with a high level of curiosity about certain classes of animals — reptilia, for instance — that not infrequently translates into deep affection. For some people, all animals are innately fascinating and have an intrinsic right to exist — and the interest of such people is not lessened because a particular animal possesses the capacity to be deadly to humans. For others, being deadly to humans actually constellates their interest in a particular species. They might start out as hunters and have a conversion experience, putting their considerable knowledge back into the conservation of their erstwhile prey. Some individuals have a compelling attraction to a single species or group of species that amounts to monomania. Depending on how they choose to act this out in the world the results can be a triumph of conservation or a tragedy for the animals, the human, or both. Finally, there are some kinds of human–deadly-animal interactions that are not remotely connected to any earthly realities of animal existence. Members of the Church of Christ with Signs Following take literally the biblical words from Mark 16:17–18 (King James Bible).

> *And these signs shall follow them that believe; In my name shall they cast out devils; they shall speak with new tongues; They shall take up serpents; and if they drink any deadly thing, it shall not hurt them; they shall lay hands on the sick, and they shall recover.*

Accordingly, the believers drape themselves with deadly snakes in defiance of the secular law of the United States. An interest in the animals themselves is not required — they are merely props for the demonstration of faith.

This book focuses mostly on rehabilitating the reputation of the usual suspects by providing information about their natural history and their place in ecology and conservation. Some of the species featured — elephants and hippopotamuses — may be surprising. Their public images do not generally focus on their fatal capacities, but those who live with or near them are well aware of the threat they can pose to human life. Of course, anything short of an encyclopaedia of dangerous animals is going to have to be very selective about the species included. Here, the basis for selection was the major groups of dangerous animals, and the various ways in which people relate to them. Along the way, however, a motley group of animals presented their credentials as actually or potentially lethal, dangerous or interesting for one reason or another. Although they did not make the final cut, it's fun to have a brief look at some of these before proceeding to the featured species.

THE RUNNERS-UP

There's an old Australian curse that runs 'May all your chooks turn into emus and kick your dunny door down' (may all your domestic poultry … kick your outside toilet door down!). Maybe it's because emus are found throughout Australia, or maybe it's hard to fit the word 'cassowary' into a curse, but for my money, this fruit-eating cousin of the emu is larger and flashier. It's also been deadly on at least one occasion. In April 1926 at Mossman, North Queensland, the unfortunate Phillip McClean paid the ultimate penalty for some youthful foolishness. The sixteen-year-old and his mates were chasing a cassowary (*Casuarius casuarius*) when the youth fell. The scimitar middle claw of the cassowary caught his jugular vein. Most cassowary–human encounters don't end in death, but the possibility exists. Especially now that ecotourism in Queensland attracts people who are looking for a close encounter with wildlife, and hand-feeding cassowaries is a feature. Tourists and locals regularly report being attacked or frightened by hissing cassowaries, and a few dogs have come off the worse for wear in close encounters. Left to their own devices, however, cassowaries are shy, non-aggressive, increasingly rare and in some areas endangered. But they do not take lightly to any perceived threats to their chicks. There are plenty of ways you can die through diseases caught from birds, but the cassowary is a most unusual, if not unique, member of the bird kingdom in that it can cause direct death.

As a born and bred Australian I'm ashamed to admit that it was only while researching this book that I found out our enchanting platypus is actually a venomous animal. Well, the male of the species, at least. Adult males have ankle spurs connected to a venom gland. Although this venom is known to be excruciating to humans, human envenomation (that is, the actual injection of venom) is extremely rare, partly because direct human–platypus encounters are not a feature of everyday life for most of us, and partly because it is thought that the point of the venom is to fight off other males and win the mating wars. The gland is known to be most active in the breeding season.

Another seemingly innocuous animal, the marmot, stands accused of being responsible for countless deaths, albeit indirectly. Rats are commonly acknowledged as the major vector for the spread of the Black Death, which in several waves of plagues devastated Europe and completely changed the path of history in that region. Reservoirs of the infection still exist in many developing countries and cause significant problems there, and even in the United States between ten and twenty people a year contract the disease. Worldwide, the last serious outbreak of bubonic plague was in India in 1994, but proper public health measures meant it was swiftly brought under control.

Marmots are an appealing, squirrelly kind of animal that look either harmless or delicious, depending on your background. In Mongolia the bobac marmot is considered a delicacy. All parts of the animal are eaten bar the armpit, which is said to house the spirit of a dead hunter. Eerie, considering that the marmot was the chief vector for the *Yersinia pestis* bacterium that causes the bubonic plague, commonly marked by hideous swelling in the armpits of both humans and marmots. *Yersinia pestis* infects the lungs of bobac marmots. When infected animals cough they spread the bug to fleas, rats and ultimately humans. Inadvertently, then, some have argued that the marmots have an impressive toll of more than one billion human deaths during the great plagues. Marmots would be an animal worth developing a decent phobia about, but they are not widely thought of outside Mongolia.

Other animal species are known to one and all but are not recognised as much of a threat. Unlike sharks, deer seldom feature in people's nightmares. Unlike mountain lions, cows are unlikely to inspire terror. Ultimately, if there is one repeated theme in this book it is that perceptions of danger are not necessarily rational and frequently bear no relation to the real dangers people are likely to confront. According to a table on the University of Florida's Shark Alert site, sharks and mountain lions accounted for fewer than one fatality a year each in the United States during the 1990s. On the other hand, deer — those innocent Bambis — topped the list of dangerous animals. On average, during the same period 130 people in the United States died due to deer each year, generally through vehicle collisions. Neither dogs, at eighteen per year, nor snakes, at sixteen, came close. Clearly, nobody is suggesting that deer are particularly aggressive, and as herbivores they are hardly classifiable as predators on humans. It is just good to put things in perspective. As the old saying goes, if you hear hoof beats think horses, not zebras.

And speaking of horses, they — along with cattle — merit attention, because a number of human deaths are attributable to both species. Dr Ricky Lee Langley and veterinarian Dr James Lee Hunter researched work-related deaths involving animals in the United States over five years from 1992 to 1997. Although animals were only a factor in 1 per cent of workplace fatalities, this still represents the loss of 350 lives. Cattle top the list at 142, closely followed by horses. Two species that aren't native to the United States — tigers and elephants — killed five and three people, respectively. Amazingly, snakes were not even mentioned separately, but probably accounted for at least some of the ten deaths listed under 'Other'. In many cases it seems that human fear of dangerous animals is in inverse proportion to the likelihood of being killed by a particular species. A bit like people who are terrified of flying yet happily drive a car!

THE SCIENCE OF NAMING

Anybody with the least interest in natural history is aware of the system of scientific classification based on the work of Linnaeus, and the frequently stated benefits of an internationally understood taxonomy. Common names are, well, common — and very often based on the most obvious feature of a plant or animal. Thus the name brown snake or Australian brown snake is frequently used to refer to any one of several highly venomous members of the *Pseudonaja* species. However, brown snake — a simply descriptive name — could easily be attached to any number of snakes anywhere in the world, some of which are harmless and some not. Alternatively, a range of completely idiosyncratic and regionally used names can all be attached to the exact same widely distributed species. It is quite clear that a properly constituted scientific system of naming is essential to clear up these confusions. And usually it does — but not always.

This is how the system works for all animals: a type specimen — called the holotype — is formally described in the published literature and is given the nod by the International Commission on Zoological Nomenclature as fulfilling the International Code of Zoological Nomenclature. The short term for taxonomic work is 'systematics' — and it is all subject to revision. At the broadest and most basic level, the taxon are defined as Kingdom, Phylum, Class, Order, Family, Genus, Species, in increasing order of specificity, with the last two together defining a single species group. Naturally, the broader the levels of classification, the more stable the category. It would take a scientific revolution of major proportions to overthrow a kingdom! Down at the pointy end, however, the genus and species classifications change with an amazing frequency. Even before the advent of molecular biology and subsequent revolutions in systematics, there was the issue of scientists in different regions or continents describing new species that were later found to be conspecific (identical) either with each other or another already long-described species. Newly discovered species are not always so new. Conversely, some well-established genera or species have been radically revised and reorganised, particularly with the new techniques available such as the analysis of mitochondrial DNA. Disputes about classification can be quite heated — and the cognoscenti laughingly refer to it as the war between the 'clumpers' and the 'splitters', depending on how species are grouped.

DANGEROUS AND DEADLY — WHAT'S THE DIFFERENCE?

Dangerous: even the word can set pulses racing, but what does it actually mean? For some, danger arouses responses that range from nervousness to blind panic; for others there's a mysterious allure mixed in with the adrenaline response. It takes all sorts, as they say. Similarly, there are all sorts of ways of looking at the idea of 'dangerous animals'. For a start, it's worth noting that the terms 'deadly' and 'dangerous' may be related, but they are not interchangeable. It is important to untangle them. An animal may possess the capacity to be extremely deadly, but if it lives alone with others of its kind on a remote island or in a remote desert, it can hardly be called dangerous.

An animal that is deadly, and lives where humans live, can rightly be described as dangerous. Even here, though, there is a strong distinction to be made between animal behaviours that are aggressive and those that are defensive. It would seem obvious that the former should be considered more dangerous than the latter, but people's perceptions are not always accurate. For example, worldwide the image of the rattlesnake is one of great dangerousness, and (unless you live near one) the hippopotamus is likely to be viewed as a wallowing old duffer; yet the rattlesnake is a gentle-natured creature, and the hippopotamus is not. Even the category of aggressive animal divides again into animals that are permanently aggressive and those that have seasons of aggressive behaviour.

Of course, almost anything can be dangerous or toxic, depending on the circumstances. A fly can distract a driver who then drives into a tree and winds up dead — the driver, not the fly. But that's pretty random, so let's take a broad look at the more predictable categories of animals that can kill or seriously harm humans, and their modus operandi. A whole range of animals, apart from the hypothetical fly, are involved in fatal car crashes. Most of them, of course, fatal for the animal, not the human. But a significant subsection of collisions causing human death involves large animals, particularly those that move quickly or unpredictably. Kangaroos in Australia, deer in the United States, and bulls wherever they may be are examples. The ubiquitous bull bar attests to this danger. Statistics clearly show that this is not a negligible risk, but even the briefest time on the back roads littered with the rotting carcasses of wildlife will show that fatal outcomes are heavily weighted towards the animal in these encounters.

KEEPING RECORDS

What do a shark, cow, bear and horse have in common? Not much, apart from being lumped together under the 'other specified animal' category of the International Classification of Disease (ICD) coding system. As researcher Ricky Lee Langley points out, this makes it hard to obtain completely accurate data about how many human deaths are attributable to specific species of dangerous animals. He recommends getting a little bit more detailed about the statistics. As the international reporting system undergoes constant revision it is necessary to keep track of the shift in how various animal-related deaths are coded from one edition to the next. One of the difficulties in gaining genuinely scientific statistics internationally is the variety, and all-too-often the inadequacy, of various methods of collecting information, particularly in countries where other issues may be of more pressing importance. Frequently data is not collected at all, and when it is, it is often collected piecemeal with different systems used by different ministries, agencies and departments, including police, legal, health and labour. In South-East Asia, for example, only Thailand uses the ICD codes. Without accurate and properly comparable statistics, it is very difficult to make absolute judgements about the relative dangerousness of various animal species. It also makes it much harder to formulate effective prevention plans based on sound epidemiological data.

MODUS OPERANDI

It's clear enough that the modus operandi of dangerous animals divide pretty neatly into physical and chemical means. It must be recognised, though, that in the vast majority of cases, humans are collateral damage — unfortunate or foolhardy enough to encounter a potentially dangerous animal at the wrong place, wrong time! Most of the animals that kill us, even those that do so with remarkable efficiency, have not evolved to specialise in doing so. Venoms are used to anaesthetise or paralyse prey species — such as amphibians and insects, though a smaller number of venomous species favour eating small mammals. (As mentioned above, countless billions of bacteria are accidental avatars of death by disease. In fact some of the nastiest hard-to-heal wounds attributed to venom in snake and spider bite victims are actually the result of extremely infectious microbes introduced whether or not the animal also injects venom.) Some small invertebrates, particularly bees and wasps, can cause fatal allergic reactions in certain individuals. Of course, some classes of venom are among the most efficient death dealers. Various species of spiders, scorpions, snakes, cone shells, stingrays and box jellyfish are all purveyors of venom.

Generally, however, in thinking about deadly animals it is direct body-to-body attack that comes to mind rather than car crashes or microbes — mauling, for example, which includes beating, biting, chewing, clawing, goring and eviscerating. To effectively achieve this on a human an animal has to be pretty big itself, so the key species here are sharks, crocodiles, big cats and bears. (Mind you, rats have been known to kill babies.) If you add crushing, stomping and suffocating to the list you can count in hippopotamuses, cattle, horses, camels, moose, elephants and giant anacondas. In February 2007 an eight-year-old Brazilian boy had a lucky escape from a 5-metre anaconda that had wrapped itself around his neck. Journalist Tales Azzoni wrote that police reports revealed Joaquim Pereira saved his grandson Mateus from the grip of the anaconda by beating the snake with rocks and hacking at it with a knife for half an hour.

Anacondas are obviously a real threat — if you live in their habitat and are vulnerable (by virtue of small stature) to their means of predation. They barely register, however, in the overall picture of deaths caused by venomous snakes or other wild animals. Often it is the commoner domesticated animals, such as dogs, that pose a more significant danger. Domesticated or otherwise, in-season males of more than one species are a danger. Robyn Davidson (whose idiosyncratic pilgrimage with camels across Australia's deserts was detailed in her book *Tracks*) has experienced this first hand. If she had any reason to doubt the Afghan cameleer's advice to 'shoot first, ask questions later' when encountering wild bull camels, she was convinced of its wisdom when her own 'baby' bull camel, crazy with his first flush of adolescent hormones, just about killed her.

Another flash point for human–animal encounters with these large species is inadvertently preventing a 'mother and child reunion'. Hell hath no fury like a mother who perceives a threat to her offspring. It is not recommended to get between a bear and her cubs — and the same applies to mothers of most other large mammal species. There are, of course, exceptions to every rule. A friend was hiking along a creek bed in the United States, looking down at his feet, concentrating on finding the next stepping stone. He failed to see what he was just about to walk into: coming the other way, also unprepared for the encounter, were a black bear and her cubs. A man of science, he knew full well that this was about as dangerous as it gets. In a moment worthy of a Larson cartoon, both man and mother bear looked at each other for an instant, equally horrified. Then in the same instant the man scaled a cliff, and the bear took to her heels and fled in the opposite direction. It was all the cubs could do to keep up with her. Lucky man; lucky bears. Never forget, though, that the exception only goes to prove the rule!

ENDANGERED

Revealingly, the word danger is embedded in the word 'endangered'. So many species that humans rightly or wrongly consider a major threat end up on the endangered and critically endangered lists. Crocodiles, sharks, snakes — the list goes on. Sometimes this is through active culling, sometimes through human-caused habitat loss. Sometimes it is a mixture of both factors, as when habitat loss causes more frequent encounters between humans and dangerous species, leading to increased calls to exterminate the animals. It is reasonable to wish to protect human life; but looking at the bigger picture gives a better balance about levels of danger and reasonable measures to deal with them. It is only fair to ask who is more dangerous to whom? An indication of the conservation status of the animals in this book will be based on the International Union for the Conservation of Nature (IUCN), as follows.

IUCN CLASSIFICATIONS OF DEGREES OF THREAT

EXTINCT: Species has not been located in the wild during the past 50 years.

ENDANGERED: Species in danger of extinction. Survival is unlikely if the cause of its decline continues.
VULNERABLE: Species believed likely to become endangered in the near future if the cause of its decline continues.
RARE: Species with small world populations that are not at present endangered or vulnerable.
THREATENED: A general term used to describe a species in one of the above categories.

PAYING OUR RESPECTS

R-E-S-P-E-C-T, so goes the song. It's a good song, a good word. One that bears a little examination because it's the key concept underlying a sustainable way of relating to dangerous cohabitants of the planet. Sometimes respect is paired with affection, but not always. There are plenty of people on the planet who are besotted with snakes, spiders, sharks and a whole range of creatures that inspire no love in others. It is neither possible nor necessary for everyone to share this infatuation. But we can strive to encourage an attitude of respect for all creatures based on their intrinsic right to be alive and thrive in our shared home on Earth. This does not automatically translate to literally sharing our intimate living space — that can be left to the seriously devoted!

When we are dealing with dangerous creatures, another aspect of the word respect comes into play: a realistic idea of the threat they pose — based on accurate knowledge — which leads to appropriate ways to reduce or avoid the risks. Ignorance and fear are bunkmates, and they frequently raise risk rather than reduce it. In the developed world, the vast majority of snakebites occur because people are persecuting, pestering or handling snakes carelessly. Even less complicit victims of snakebite have often failed to take simple, commonsense precautions based on knowledge of local snake behaviour.

Deadly Beautiful is dedicated to promoting respect for dangerous animals in both senses of the word. These creatures that loom so large in schlock-horror nightmares are frequently hunted to the edge of extinction and beyond. Border disputes can be dealt with very effectively without disrespecting the right of all animals to exist. The knowledge of real risks is presented here to encourage respect and better ways of learning to live together.

THE DAMASCUS AWARDS

Most people have heard of the Darwin Awards — given out, tongue in cheek, to those whose blatant foolishness contributed to their own death, thus 'removing their genes from the pool'. Not surprisingly, those who mishandle or bait dangerous animals are frequently nominees. Grimly amusing, perhaps, but not very cheerful. The Damascus Awards, on the other hand, are evidence of humanity's capacity to change and grow, and provide a far more positive view. No gold attaches to winning a Damascus Award, only the satisfaction of giving something back. Invented for this book, the award is of course named for St Paul's conversion experience on the road to Damascus. It is given to honour those people who started out life as hunters — and sometimes active persecutors — of dangerous species, and ended up using their knowledge to fight for the preservation of their erstwhile enemy. Several chapters of *Deadly Beautiful* feature at least one winner of a Damascus Award.

1

A Matter of Scale

Snakes

The crèche is well attended. Located in a rural community in North America, it exudes a sense of stability and peace. There is a real comfort in the familiarity of generations of mothers who have gathered here to share the warmth and undergo the trials of pregnancy and childbirth in the company of others. Suddenly, the peace is shattered by the threatening rumble of large vehicles, the overpowering smell of diesel exhaust. The men arrive with guns and clubs, crowbars and gasoline. When they are done, the only thing left is a few broken rattles. It is not the first time, but it may be the last. The community cannot hold out much longer.

Timber rattlesnakes are creatures of tradition. Traditions of their own devising include the habit of pregnant snakes coming together to sun on basking rocks at the same location every year. In winter, both sexes will den together somewhere free of frost, but not too dry. Always the same somewhere. These dens are called hibernaculum: a winter tabernacle of snakes! The latest theory for why this occurs is that it is more motivated by mate meeting than the need for heat — but perhaps the former provides the latter. Denning is more likely to occur in the colder northern parts of North America, so the jury is out on this one. Native American people called the snakes 'Grandfather', a term of great respect. When the colonists came, they too had a poetic turn of phrase, referring to the rattlesnakes as 'belled vipers'. But there was no respect. Appalachian snake dens, in particular, evoked a peculiar horror in the puritanical mind with its central association of snakes with evil. They are easy prey, those mothers of habit, those grandfathers of tradition. So, year by year, the men come (it is mostly, although not exclusively, men) and state by state the timber rattlesnake falls extinct.

Anyone who has bothered to spend time really observing the timber rattlesnakes — the local Native American people, the herpetologists and naturalists — uniformly reports that they are gentle, non-aggressive and will avoid rather than pursue humans. Sometimes, even those who have actively hunted the species slow down enough to recognise that the aggression originates on the human side of the equation (see the Damascus Awards at the end of this chapter). Some researchers have gone well beyond the call of duty to provide evidence of this lack of aggression in certain species, up to and including standing adjacent to rattlesnakes and placing a foot (softly) on a rattlesnake's back, leaving enough 'play' in the snake's length for it to bite if it chooses. Not a course I would recommend for several reasons, but nonetheless it has provided some data on the relative lack of aggression in the species thus studied. Rattlesnake data is particularly scarce; one might hypothesise that this is because scientific evidence is likely to dispel the myths held so dear by some about the level of danger posed by rattlesnakes.

SNAKE BASICS

There's more actual data about many other types of snakes, but it's hard to untangle. One of the first things to make clear is that snake research comes from two distinct areas of science, with different aims and agendas. Those who study the snakes' biology, habitat, ecology and natural history are called herpetologists. It is a field comprising both professional zoologists and amateurs, in that strict sense of the word: these people are true 'lovers' of snakes as well as other reptiles and amphibians. It would be a given, presumably, that this group is committed to the conservation of the species they are studying. Another group of scientists with research interest in snakes are the medical toxicologists. Their interest is in the epidemiology of snakebite, the composition of venoms, the clinical effects of snakebite, the development of antivenoms, and the potential use of venom or venom components in other areas of medicine. Despite close encounters with the distressing and sometimes hideous effects of untreated snakebite, none of this group appears to be anti-conservationist — many combine their research interest in toxicology with an affectionate fascination for snakes. The efforts of toxicologists are dedicated to reducing the global impact of snakebite by eliminating ignorance and inequity rather than eliminating snakes.

As well as generally being disposed towards the conservation of snakes, each group of researchers is in complete agreement about some basic facts. Snakes are not slippery. Neither are they slimy. Nor, for the most part, are they aggressive. Snakes have a single lung. They do not possess eyelids, legs (apart from vestigial lumps in a few species) or external ears. This does not mean they are deaf, simply that the mechanism of their hearing is somewhat different. The sound waves travel through the skin and muscle to hit the bone underneath. Although they can hear, it is not their best sense, and it works well only in the lower ranges of vibration.

In terms of smell, however, snakes excel. They do have forked tongues (literally rather than metaphorically!) that 'read the air' for a range of scents. They can all eat improbably large meals by virtue of being able to unhinge their jaws. This means that sizeable and infrequent meals are common in snakes' dietary patterns. They are top of the charts with regard to vertebrae (120 or so is the bare minimum for any snake species), hence that sinuousness, and the amazing ability of some species to coil. The well-endowed *Morelia oenpelliensis* has 585 vertebrae, compared to our measly 33. Less than 20 per cent of all snake species are venomous. Despite this, snakes are the subject of a disproportionate amount of urban and rural legends. Beyond these acknowledged facts, the professional diversity of opinions and approaches not only between herpetologists and toxicologists but also within each group would startle the non-scientist.

MEET THE FAMILIES

Four serpent families encompass the world's most venomous snakes: Colubridae, Elapidae (which includes sea snakes, subfamily Hydrophiinae), Viperidae and Atractaspididae. There are at least 2600 species of snakes in the world, with another 2000 subspecies. However, although 1700 of the 2600 snakes belong to the largest family, Colubridae or colubrids, only a very small proportion of this group are venomous, and even fewer are dangerous to humans. One primary feature of the colubrids is the presence of fangs in the back of their jaw. The African boomslang is an example of this. The elapids, on the other hand, excel at being venomous and although many of them do not pose a threat to humans a significant minority of species do. This group includes cobras, king cobras, kraits, 'brown snakes', 'black snakes', taipans, death adders, tiger snakes, copperheads, coral snakes and sea snakes. As well as being venomous, elapids all have short fangs that are permanently in an erect position and, unlike those of their colubrid relatives, their fangs are located in the front of their jaws.

In contrast, the long, hinged fangs of the vipers are normally tucked up against the upper jaw until required. The viper family is split into two groups: typical vipers and pit vipers. It is easy to assume that the latter name is related to the cliché 'snake pit', but it is not. Pit vipers do not reside in pits. In fact, they can be arboreal (live in trees). The name refers to a special organ used to sense the nearness of warm-blooded prey. It looks like a small dent midway between the eye and the nostril. Trust me on this one: eyeballing a live viper at the right range to verify this information would not be a good idea.

Atractaspidids are the new kids on the block, taxonomically speaking. Formerly classified as either colubrids or vipers, the mole vipers, stiletto snakes and burrowing asps are now considered a legitimate group in their own right. Not many of them are a major problem to humans, though, with only three species known to have caused fatalities.

HOW TO HARASS A HERPETOLOGIST

If you want to annoy a herpetologist, just call reptiles 'cold-blooded' — the correct term is exothermic. This is more than mere pedantry. Snakes, like all other reptiles, generally need to keep their blood temperature within a reasonable range (i.e. warm); it's just that they achieve this by different mechanisms. All that basking in bushes and on rocks and roads is precisely in order to avoid cold blood. And the term 'cold-blooded' is freighted with so much negative meaning for humans. We call the worst of human murderers 'cold-blooded', and the general sense of distaste associated with the words spills over to the animal group we describe in the same terms.

And while we are on the subject of herpetologists' pet peeves: poisonous and venomous are not synonymous. Poisonous refers to anything, plant or animal, that is toxic either when you eat it or come into contact with it. Venomous, on the other hand, refers specifically to animals that are equipped in one of several ways (spines, fangs, etc.) to puncture the skin and thereby introduce substances of varying levels of toxicity to the body, either to protect themselves or as a feeding mechanism. Although one of the major functions of venom is to immobilise prey that would otherwise escape, venom is simply modified saliva that serves to begin the process of digestion; after all, snakes cannot chew. Apparently, most venoms are harmless if ingested. I'll take that on trust, as I'm not willing to provide empirical evidence by swallowing some.

SNAKES ARE NOT SLIPPERY, STATISTICS ARE

All the experts agree that the first, and for decades the only, definitive work on how many people worldwide die from snakebite was written by Swaroop and Grab in 1954. Their research covered the 40-year period up to 1954. Although their work is highly regarded as the first and best attempt to collect such comprehensive data, current experts also agree that the 1954 figures are far lower than the actual figures for international snakebite deaths, perhaps by as much as 40 per cent. The most recent global statistics were published in *Nature Biotechnology* in 2007. Roberto Stock and fellow researchers extrapolated from a range of hospital records and other official sources to reveal figures showing a marked discrepancy between the developed and developing world, reflected in both the number of snakebites and the mortality rates.

AROUND THE WORLD IN 80 SNAKES

In Europe, the only dangerous snakes are four species of vipers. The common adder reigns as the most deadly snake, but only around one-third of those bitten by any of the vipers are actually envenomed, and of those less than 0.5 per cent die. Despite the hysteria about rattlesnakes (granted, they are the most dangerous snakes in the United States) less than one-sixth of all people bitten by a snake *of any species* in North America are envenomed.

(There is some evidence to suggest that in up to 80 per cent of rattlesnake strikes, the animal keeps its mouth shut.) And the death rate is the lowest in the world — less than 0.25 per cent. This tiny figure must include those whose faith in religious snake handling precluded seeking medical assistance, so hypothetically the preventable death rate must be lower still.

Asia's most dangerous snakes — cobras, kraits and Russell's vipers — take a major toll. The Asian region has almost double the number of snakebites of all other regions of the world put together. It has four times more snakebites than Africa, but both Asia and Africa share exactly the same rate of envenomation (50 per cent) and a very similar death toll (5 per cent of all bites for Asia and 4 per cent for Africa). Although Africa has to contend with the extremely dangerous puff adders and spitting cobras, it is the vipers that are the worst. According to the authors of the article in *Nature Biotechnology*, the ocellated carpet viper (*Echis ocellatus*) is the single most dangerous species in Africa, accounting for more than 70 per cent of snakebite deaths. Oceania has the smallest human population, the lowest figures for snakebite, and only one-third of those bitten are envenomed. But those who are have the highest death rate in the world (nearly 7 per cent). That is perhaps not surprising given that the deadliest snakes in Oceania are the eastern and western taipans.

All researchers acknowledge that the epidemiological information is far from complete. However, in the absence of centrally collected, uniformly coded global statistics, this is as close to an accurate picture as we are going to get of the real death toll. There are so many complicating factors. Just for a start, herpetologists frequently make an informal but clear distinction between 'real' snakebites and 'stupid' or 'illegitimate' snakebites. And they are not attributing stupidity or bastardry to the snakes! What they mean is that, particularly in the developed world, a fair number of avoidable snakebites are caused by the interaction of two chemicals, neither of them in venom. Testosterone and alcohol fuel an enormous number of risk-taking behaviours, including being foolhardy around snakes. Just as dangerous, but perhaps more understandable, are those who feel obliged to catch and kill a snake for identification purposes after or even before a snake has bitten. This is neither necessary nor advisable.

Obviously, working with snakes or keeping them as pets greatly enhances your chances of being bitten. The amateur herpetologist who first 'rediscovered' the western taipan actually thought it was an Australian brown snake (also venomous), which looks very similar. He was fortunate enough to survive being bitten. But not everyone is so lucky. Even the most experienced and respected herpetologists are at risk. It just takes a split second to make the wrong decision around deadly snakes, and the outcome can be fatal. This was the case for Dr Joe Slowinski a renowned herpetologist from the Californian Academy of Sciences. The 39-year-old died after being bitten by a multi-banded krait (*Bungarus multicintus*) in northern Burma, far from medical aid. Because it had allegedly bitten one of the Burmese members of the expedition the day before, Dr Slowinski felt confident that he was picking up a non-toxic, almost identical snake, but he was instantly aware of his mistake. Unfortunately, it proved to be too late.

As well as distinguishing between 'real' bites and 'stupid' (i.e. avoidable) bites, there is also the category of 'dry' bites, where no venom at all is injected into the victim. In some species as many as 50 per cent of snakebites are dry. Anaphylactic shock is a known cause of death in envenomed snakebite victims; perhaps it is remotely possible that some people who have suffered a dry bite might actually die of fright. Herpetologists fairly claim that if you correct any regional statistics by eliminating 'stupid' bites, then the threats posed to human life by even the most deadly snakes are very low indeed. Again, it's a relative matter. No one is arguing such snakes pose zero risk. No one is claiming that 'stupid' bites aren't deadly. More to the point is the fact that most venomous snakes living out their lives undisturbed in their natural habitat are very rarely the cause of human deaths.

Obviously, there are a couple of caveats on the above statement. 'Undisturbed' is one and 'natural habitat' is another. One of the most venomous snakes in Australia is the western taipan. It lives in the remote outback and the drainage areas of the legendary Diamantina and Georgina rivers and Cooper Creek. The human population there has always been very low, and is lower still since European settlement disturbed the traditional patterns of the indigenous people. Aboriginal people in the area have had a long time to work out their cohabitation with the local fauna. Consequently, the western taipan may be the most venomous snake in Australia but it can hardly be classified as the most dangerous (see 'And the winner is ...' on page 37 for which snake is the most dangerous and why).

'Natural habitat' is the major issue. In some regions of the world, such as India, the large populations of rural dwellers are well accustomed to living with wildlife. This is a matter of pragmatism, and can lead to strange encounters. One resident of an Indian village was asleep with her two children when she was bitten by a cobra. Acting on who knows what reflex — perhaps to protect her children — she promptly caught the snake and bit back. After being treated at the local hospital she survived. To paraphrase Oliver Goldsmith, 'The snake it was that died'! The woman was fortunate. Although facing serious issues of population and poverty, India is highly advanced in the organised approach to treating snakebite. Unlike most developing nations, it possesses a national reference collection of antivenoms.

VARIOUS CONTENDERS

There are no serious international statistics kept for snakebite mortality. Because of this, various countries are able to claim the dubious honour of being 'Snakebite Capital of the World'. Sri Lanka, Nepal and Papua New Guinea are all contenders for the title, but until a formal central register is kept no one will really know.

GETTING ON WITH THE NEIGHBOURS

Developed nations, with dense urban populations, do not have a good record of habitat-sharing. City dwellers may come to an uneasy truce with rats, cockroaches and pigeons but they are not normally confronted with wild snakes, unless they are near remnant patches of undeveloped land. For several years I worked at an urban Western Australian university that listed a 'snake catcher' in its internal phone directory. In spring it was not uncommon to encounter the local dugites (*Pseudonaja affinis*) sunning themselves around the grounds, and one was once found on the fourth floor of a university building. Did it drop from a huge gum tree onto a window ledge and enter through an open window? Did it take the lift, like a denizen of a Larson cartoon? Did it slither unseen up four flights of stairs? No one knew. But the snake catcher was called, the snake safely returned to its habitat, and academic work resumed. Dugites are dangerous, and the university community is annually warned of the snake activity in spring when the mating season may make them more aggressive and of the emergence of venomous juveniles in late summer and throughout autumn. But there is a reasonably cordial cohabitation.

WHAT'S IN A NAME?

Is the proper term antivenom, antivenene or antivenin? Do they mean different things? No — actually, the terms are synonymous, with antivenin being considered an old-fashioned nineteenth-century version of antivenene, and antivenom merely being a seemingly more recent variant.

ASSESSING THE RISK

Cordial cohabitation is easier when you can realistically assess a situation. Neither size, nor pattern, nor location (of the snake or the bite) is an infallible guide to the risk you face in any particular case of snakebite. Even when a venomous snake is dead you are not 100 per cent safe. There are a number of documented cases of snakebite inflicted by 'dead snakes' — sometimes, gruesomely, by the severed head of a dead snake. Rarely, even the cast-off fangs and skins of some species have caused adverse reactions, although, despite wild rumours to the contrary, never human death.

It is definitely possible to be bitten by a snake's head for up to an hour after the act of decapitation. This was safely proven (at least from the human point of view) by a researcher called Klauber in 1956 who decapitated rattlesnakes in the interest of science (!) and found they could indeed reflexively bite allegedly post-mortem. An unfortunate Florida man found out the truth of this the hard way when he was bitten by the detached head of the timber rattlesnake he was preparing to cook up and eat. He never got to have his last meal. This is the stuff of horror movies, both for man and snake. Decapitated snakes are dying but not actually dead. As a result of a reptile's metabolism and tolerance of low levels of oxygen (anoxia), a snake's head detached from its body is actually fully conscious and its brain is functioning perfectly since it doesn't require high levels of oxygen. Therefore, decapitation in *all* reptiles is inhumane. Crocodiles, for example, may be fully conscious to pain for up to two hours after they have been so-called killed and bled. So the old myth that killed snakes do not die until twilight actually has some scientific truth encased in its poetry.

The pattern of snakes' skins is often very deceptive. It's confusing enough to scientists, who frequently have to revise their classifications, but to a lay person it can be downright confounding. That's what makes snakes such masters of camouflage. There's a reason the South American bushmaster came by its name. A friend who is a highly observant scientist once radio-tracked a western diamondback to within a couple of metres, and still could not see it lying in front of him. Trying to identify the species of a snake I once saw at my brother's property in rural Australia, I looked up a website that had several pictures of radically different pattern variations for a single species. So the pattern alone made me none the wiser about the correct identification.

In terms of snakes, size is not the most pertinent factor. Big snakes might look scarier but some of the small vipers are just as deadly, if not more so, than larger venomous snakes; for instance, the highly venomous saw-scaled viper (*Echis carinatus*), one of India's most dangerous snakes, seldom exceeds 60 centimetres in length. Size does matter insofar as the body weight of the victim (children are at far greater risk from venom than adults) and for that reason antivenom, unlike other forms of pharmacological treatment, is not scaled down. A child should be treated with the full amount of antivenom that would be administered to an adult.

Location counts if you live on the following islands: Hawaii, Iceland, Ireland or New Zealand. These are places known to be entirely free of venomous serpents. Other than that, venomous snakes of varying species live in every continent (apart from Antarctica). Bite location also counts if venom is injected directly into your vein by a bite from a particularly venomous species, known colloquially in the United States as a 'two-stepper'. In other words, you take two steps and die. The universality of this fear is evident when you consider that in Somalia they use the expression seven-stepper — as well as 'father of ten minutes' — to denote the same idea (they must be hardier in Somalia!). Another apt African expression translates as the fatalistic 'Do not bother the doctor'. Similar claims were made in letters to the Mother Country from colonial Australians. These phrases can all be taken with a pinch of salt. Apart from exceptional circumstances, death from untreated snakebite, even from the most deadly species, is usually a matter of several hours at least. Also, you would have to be very unlucky for the venom to be directly injected into the circulatory system. In Australia, the venom of elapids travels via the lymphatic system; consequently, a broad bandage, wrapping the entire limb of a bitten extremity, is effective in slowing the movement of venom through the body. By far the majority of snakebites occur on the extremities — hands and wrists or feet and ankles — and only puncture the skin, not the veins.

AND THE WINNER IS . . .

When it comes to ranking the world's most deadly snakes several factors need to be assessed before it's possible to realistically judge the threat a particular species poses to humans. It's a bit like detective work — a snake needs to have the motivation, the means and the opportunity. As no species of snake has evolved to prey on humans for food, hunger cannot be counted as 'motivation'. The relevant motivational factors are defence and aggression. There is general agreement among herpetologists that the majority of snake species are not aggressive, and that on the whole snakes are much more afraid of you than you are of them. And with good reason.

The Buddhist calm of most snakes notwithstanding, some species, by virtue of preferring to prey on mammals (other than humans) and general irritability of temperament could qualify as having the 'motive' to harm humans. Nothing personal, you understand, just a general disposition to bite first, and possibly repeatedly. As with humans, there is a great deal of variation from one individual to the next, and even from one group to the next. The consensus among people who deal with snakes professionally, however, is that one of the frontrunners in the 'irritable snake' stakes is the taipan.

Still, there would be no point in fearing an aggressive species unless it had sufficient means to kill you: it would be all bite and no venom. Venoms are extraordinarily complex compounds, with multiple and not always predictable effects. These effects can be very broadly divided into two major groupings: haemotoxic (primarily affecting blood and tissues) and neurotoxic (primarily affecting the nervous system), but some venoms in some cases will produce both effects. So the toxicity of the venom, the ability to inject a large quantity of venom, particularly the capacity to bite repeatedly, all comprise the 'means' for a snake to kill humans.

That leaves 'opportunity'. Translated that means asking the question 'Does the snake live near large human populations?' A snake can be as aggressive and venomous as it likes, but if it seldom encounters humans it hardly constitutes a real danger. So, which species is going to slither up the red carpet when that envelope is opened to announce the winner of the world's most dangerous snake? Which country has the honour of being home to this species? According to Australian researcher Jeanette Covacevich, it's the eastern taipan, resident of Australia and New Guinea. And she should know. She is the person who conceptualised the various factors that need assessing before assigning any ranking to dangerous snakes, a framework that is now widely adopted. She was also there the day the western taipan was rediscovered for Western science in 1974. In terms of venom toxicity, the eastern taipan is eclipsed by its cousin the western taipan, which holds the singular honour of being the world's most venomous terrestrial snake. But because the western taipan lives in inland Australia, a long way from human settlements, it is seldom involved in human fatalities. The eastern taipan produces more venom than almost any other species (its yield is only exceeded by the mulga snake), is aggressive and lives near people. It has the motive, means and opportunity. And it has the X-factor — unlike the majority of Australian snakes, its fangs are long, at up to 10 millimetres. And in this instance, size does matter.

CURING THE COMPLAINT

Three things radically improve your chances of surviving envenomation. The first two are obvious: antivenom and access to trained medical personnel. The third is the means of accessing the first two — planes and boats and trains and motorcycles and burros, horses and alpacas for that matter. Mortality from snakebites in Australia plummeted once reliable supplies of effective antivenom were available. Unfortunately, the production of antivenom is a costly business at the best of times and it's even harder to get it right all of the time. Early antivenoms in the developed nations were infamous for the frequency with which they caused allergic reactions; sometimes the cure was more catastrophic than the complaint. Although this problem lingers long in the medical memory, causing many to be conservative in the administration of antivenom, it is no longer a major issue for several reasons.

For a start, as soon as it was evident that using antivenom raised the risk of complications such as anaphylactic shock, protocols were put in place, including precautionary medication as well as having all necessary procedures, medications and equipment assembled for instant action if required. And the antivenom itself continued to be refined until it contained fewer and fewer of the whole proteins that are the major culprit in provoking allergic reactions. Some physicians with an expertise in toxicology are mildly concerned that the bad old days of rough antivenom might mean there are cases when they are withheld or victims are under-dosed, but this does not seem to be a really major issue. In the context of a well-run clinical setting, conservative treatment is fine anyway, as a good percentage of cases are resolved just by treating symptoms (pain, hypertension, etc.) and keeping a close eye on the patient.

So far, so good. We have been talking about the developed world, where your chances of being bitten are not high in the first place, and your chances of dying from envenomation are way below the chances of being struck by lightning. This, however, is far from being the case in the developing world. The figures here are sobering — both in terms of number of bites and resulting deaths. The reasons for this are evident. For a start, snakebite is classifiable as an occupational hazard in many parts of the world — it primarily affects rural workers, farmers, foresters and fishers. The combination of chronic poverty and traditional lifestyles contributes to the risk: if you sleep on the floor of a thatched hut, naturally you are more at risk of encountering the snakes who seek refuge in your roof or hunt the rodents that run in and out of the chinks in your walls. Often, sleeping people who are bitten by nocturnal snakes such as kraits do not even know they have been bitten until they become ill.

If poverty and tradition increase the likelihood of being bitten in the first place, they certainly don't improve your chances once you have been bitten. The victims and their families seldom have the resources to pay for medical treatment, which is frequently unavailable anyway. Predictably, they have recourse to traditional healers — and, sadly, many deaths have occurred because effective treatment was delayed unnecessarily when traditional treatment was the first recourse. Lest you feel tempted to indulge in a bit of smug complacency about 'superstition', there are a few things to think about.

The common English expression 'snake oil salesman' testifies to the fact that it was not so long ago, before the widespread availability of antivenom, that the populace of the West was willing to believe and purchase any number of remedies against snakebite. Secondly, obtaining empirical evidence of effective cures by traditional means has frequently been confounded by the fact that varying percentages of snakebite are 'dry' — thus lending false credibility to a cure for something that hadn't really happened — and this is the same the world over. Thirdly, traditional pharmacopoeias do contain some valid remedies. After all, aspirin can be traced to willow bark, penicillin to mould, and digitalis to foxgloves. Traditional medicine can also be harmful, either actively dangerous or passively dangerous, because even when harmless in itself it prevents or delays better treatment. On the other hand, many scientific studies are now undertaken to assess the efficacy of traditional medicine for snakebite. Ethnopharmacology is the name of a whole field of research dedicated to investigating the benefits or otherwise of traditional medicine.

COUNTING THE COST

Antivenom is very expensive. Therefore, even if a snakebite victim in a developing nation makes it to a medical facility, there is no guarantee that there will be any stock of antivenom, let alone other supplies and equipment considered basic elsewhere. For example, the African continent — which has one of the world's highest mortalities from snakebite — also has critically low stocks of antivenom, a dramatic reduction since the 1970s as African nations have wrestled with the grim social cost and loss of infrastructure resulting from the AIDS pandemic, widespread conflict and environmental disasters. And what antivenom they do have is not necessarily of reliable quality. The frequent absence of what is called a 'cold chain' (where transport and storage can guarantee the antivenom is kept at a continuously low temperature) means that 'freeze dried' antivenom is the only feasible option, and this is more problematic than liquid antivenom. It's a similar story in many parts of the developing world. The enormous loss of life when remedies are easily available is scandalous, but in the unadorned words of peace corps doctor Ellen Einterz: 'An action plan expresses no outrage'.

The horror of such situations only really comes home to you when you see it for yourself. This was the experience of Australian Dr Kenneth Winkel who spent some time in Papua New Guinea. In an Internet article published by the *Medical Journal of Australia*, he gave an alarming account of a twelve-year-old local boy who had been bitten by a snake. After the physician tried every trick in the book to get the pharmacy staff to hand over the goods they might have been stockpiling for an 'emergency', it finally became clear that there really was no polyvalent (or general) antivenom available. In such dire straits, he even considered using the two ampoules of sea snake antivenom that were still available precisely because they had never been, and were never likely to be, required. The physician eventually managed to save the child's life by dint of begging the relatives of one of the adults in the intensive care unit to give up a mechanical ventilator for the child, and by sign-language conveyed how they must continue to ventilate their own loved one by continually squeezing a hand pump. This is frontier medicine, and very often it doesn't have such a happy outcome.

As always, the best means of turning around this situation will be a combination of global and local efforts. At the global level, a number of toxicologists have been working hard to get snakebite mortality in developing nations recognised as the serious problem it is. Some of them are hopeful that a recent international vaccination program funded by the Bill and Melinda Gates Foundation can provide an infrastructure that can also be used to distribute antivenom and train local health workers in simple, practical measures to reduce deaths. Venom specialists know that the two-stepper is really a rare-to-mythical beast. That's why Australia's famous Royal Flying Doctor Service doesn't stock antivenom at any of the twenty caches of vital supplies it maintains throughout remote rural communities. There is almost always time to fly snakebite victims to a major medical facility.

Clearly, motorised transport is the second most important development for snakebite survival after antivenom itself. The in-joke among toxicologists is to give a presentation on snakebite first aid, and then show a slide of a set of car keys. In the same spirit as the Royal Flying Doctor Service, there is now an effort to organise a network of motorbike owners in mountainous and snaky Nepal so that snakebite victims can be swiftly transported to clinics. All of which goes to show that extraordinary measures are not required to meet the needs of both humans and serpents to survive and thrive. We have the means to reduce snakebite mortality in the developing world to figures equivalent to those in the West — far less than 1 per cent of those bitten. The means are simple and available. If we choose to do so, we can.

Herpetology is not everyone's bag, and it's not necessary for all of us to suddenly acquire a love of snakes. Just a little curiosity and caution in equal measures, however, would go a long way towards preventing the senseless destruction of both snakes and their habitats.

THE DAMASCUS AWARDS

Winner of the inaugural Damascus Award is Curt Brennan, author of *Rattler Tales from North Central Pennsylvania*. In his former life, Brennan won prizes for snake bagging at the infamous rattlesnake hunts so popular in the United States. Over the years he witnessed first-hand the disappearance of this species from many of its former haunts. A good observer and an honest man, he eventually found it impossible to keep promoting the false idea that rattlesnakes were aggressive, evil creatures that deserved to be hunted to death. In Brennan's words:

> *This is basically a placid and peaceful animal [...] I can't stand being a fake, and I was helping perpetuate a lie by promoting myself as a rough, tough saver of mankind by rounding up rattlesnakes. And I knew that the snake population was in trouble if I, and others like me, continued to do what we were doing. Just by getting close to the animal, and being around it for a very short time, you learn that they're not what people have led you to believe they are. If you give yourself two minutes to calm down, you'll learn the animal isn't out to get you — it's out to get away from you.*

2

HERE BE DRAGONS

Crocodiles, alligators and Komodo dragons

COURAGEOUS, CAVALIER, CARICATURE, CONSERVATIONIST, CLICHÉ: when it comes to crocodiles, Australia seems to have cornered the market on the 'C' characters. The whole Western world recognises two larger-than-life Australians indelibly identified with crocodiles — Mick 'Crocodile' Dundee and the late Steve

Irwin, whose real life exploits made fictional Mick seem somewhat tame. Crocodile Dundee did a lot to keep the Australian stereotype alive long after it was time for graceful retirement, but nobody could claim he did a lot for crocodiles. Steve Irwin, consummate showman that he was, was also more than happy to trade on extreme blokey Australianisms. Although his persona was a marketer's dream, it certainly wasn't to everyone's taste. But, crikey, for all his faults you can't deny that he had a genuine interest in crocodiles in particular and a long-term commitment to conservation in general, after his own fashion.

But the whole Western world is not the whole world, after all. And there is more to crocodilians than crocodiles. Specifically, the term crocodilian is a taxonomic grouping of 23 species broadly divided into three groups. Firstly, there is the lone Indian gharial, sole member of the crocodilian family Gavialidae. Alligatoridae is the family name for eight species of alligators and caimans, smaller snapping cousins of the Crocodylidae (true crocodiles) of which, surprisingly perhaps, there are fourteen different species. Most people are only acquainted with the two largest — the saltwater or, more correctly, estuarine crocodile (*Crocodylus porosus*) and the Nile crocodile (*Crocodylus niloticus*). Perhaps this is down to the fact that only these two really rate as serious threats to humans. On the other hand, it's also fair comment that they are in their different ways the most commonly encountered species. The estuarine crocodile has the larger distribution as it is found in the greatest range of places in the world, but the Nile crocodile, native to Africa and Madagascar, inhabits the greater area of land.

Being — like all reptiles — exothermic, it is predictable that crocodilians are generally a tropical group, present in 91 countries and islands throughout the warmer central latitudes. This is also where human populations tend to congregate, so most of us have crocodilian neighbours of one sort or another. The Chinese and American alligators can withstand a bit of cold, but they are the exception. Metaphorically, though, the Chinese alligator has been left so far out in the cold there are grave fears for the species' survival. They are an extreme case, but numbers of many other crocodilian species suffered severe depredations (estimates of up to 77 per cent falls in population) during the 'Century of Wars', which was also the 'Century of Crocodile Hunting'. Millions of animals were slaughtered and their skins used for luggage and handbags. The primary research studies into the Nile crocodile — written up in a lavishly illustrated book entitled *Eyelids of Morning* — were carried out during the 1960s. It seems outlandish in current terms, but the researchers actually funded the program by selling the hides of the crocodiles under study. Let's hope we never return to that as a research-funding model.

WHAT'S IN A NAME?

A pebble worm sounds like a harmless sort of critter to modern ears. The name of something small, ordinary, lowly even. But as readers of fantasy or mediaeval literature will recognise 'worm' is frequently synonymous with 'dragon', so that makes a little more sense of the etymology of the word crocodile. The name is derived from the Greek *kroko-drilos*, literally 'pebble-worm'. This refers not so much to their habit of living in low places but more to the pebble-like pattern of their scales.

Subsequent generations have had less of a taste for scientific understatement, perhaps, but they do cut straight to the heart of human reactions to crocodilians. You can't beat the honest economy of 'terrible crocodile' (*Dinosuchus neivensis*) and 'fearsome crocodile' (*Phobosuchus* spp.) for prehistoric crocodilians. These days, though, people seem content to let the creature speak for itself, so to speak, and the commonest common names are simply geographic locations. Thus we have the Nile crocodile (*Crocodylus niloticus*), the American crocodile (*Crocodylus acutus*), the New Guinea crocodile (*Crocodylus novaeguineae*), the Philippine crocodile (*Crocodylus mindorensis*), the Siamese crocodile (*Crocodylus siamensis*), the Cuban crocodile (*Crocodylus rhombifer*), the American alligator (*Alligator mississippiensis*), the Chinese alligator (*Alligator sinensis*) or the even more low-key name, saltwater or estuarine crocodile (*Crocodylus porosus*), which is found across the hotter regions of the southern hemisphere from India, throughout South-East Asia to Australia and beyond to Vanuatu.

MATTERS OF SIZE

In terms of size, the estuarine crocodile is the alpha and omega of the crocodilians. Like all members of the order, the males are larger than the females. Hatchlings are small, around 30 centimetres, but it doesn't take long for them to grow. For the first ten years they have a phenomenal growth rate, sometimes doubling and even tripling in size over a year. As with most living things the overall size is determined by a combination of genetic and environmental factors. Although the animals do keep growing throughout adulthood, the rate slows down considerably after they reach breeding age, which is around twelve years for females and seventeen for males. At maturity a male crocodile would easily reach 5 metres in length and weigh 450 kilograms, but maximum lengths of 6 metres are rare although more easily verified today. Weights of 1000 kilograms and lengths of 7 metres for male saltwater crocodiles were commonly claimed in recent historical times by travellers and crocodile hunters. Of course, both bravado and fear are great magnifiers. And, as with all such anecdotes, it is good to take them with a pinch of salt.

Although the jury is out on just how big the saltwater crocodile can get, it is clearly a formidable animal even at an average size but one that was nearly driven to extinction nonetheless. Worldwide populations plummeted to a fraction of their former numbers, until an increasing awareness of conservation created changes in policy and legislations. For example, in 1970 Western Australia became the first state in

Australia to declare the crocodile a protected species. Before then, something like 300,000 animals were killed, mostly for their hides, but sometimes due to habitat loss or random acts of hunting. This figure is close to the upper estimate of the current total population figures worldwide. Of those, today's saltwater crocodile population in Australia's Kimberley region is probably between 4000 and 5000. The general numbers of estuarine crocodiles are considered sustainable in Australia and New Guinea, even if they remain a bit low elsewhere in the world.

Individual estuarine crocodiles (*Crocodylus porosus*) may live to the biblical allotment of 'three score years and ten' (70 years), but the species itself has been around since long before the biblical era. Compared with sharks, crocodiles are relative newcomers. But in terms of human measures these two are real old-timers. Both evolved a winning style of survival way before the dawn of humankind (170 million years ago for crocodilians, 400 million for sharks) and ever since have had no need to improve on their perfection as apex predators. This did not grant them immunity from the Johnny-come-lately mammalian that has so successfully occupied all likely and unlikely habitats on the planet. Just recently, however, this mammalian has at last recognised that apex predators function as 'indicator species'. Because of their place at the top of the food pyramid, it takes a wide and broad base of other living species to support apex predators. Therefore, where apex predators are thriving, the rest of the ecosystem is automatically assumed to be robust.

BITER BEWARE

Like sharks, crocodiles are remarkable for the strength of their immune system — and the powerful consequences of their respective bites. All sorts of figures are solemnly quoted and repeated to 'prove' that in the latter department crocodiles win hands down over white sharks. Turns out, however, that these figures are not as straightforward as they first seem. Shark expert Rick Aidan Martin has been all too frequently asked for the definitive 'Trivial Pursuit' answer to the question of whether the white shark or the crocodile has the strongest bite. Lucidly and patiently he will clarify the complexity of deriving such comparative figures, Snodgrass's invention of the gnath-odynamometer ('gnath' as in gnashing of teeth) notwithstanding. Aidan maintains that we really only have accurate figures for the bite force of one species: human beings. He certainly doesn't deny the capacity of crocodile bites to inflict damage, but argues that the sheer length of their jaws dissipates the relative force of the bite.

So that accounts for the seemingly death-defying stunts of crocodile handlers who creep up on a croc and subdue it by holding those fearsome jaws closed with just their bare hands. Like all such stunts, there is a scientific truth behind the illusion. The musculature of a crocodile's jaw is all designed for the downward bite — only the smallest amount of muscle power is needed to actually open the jaw. This presents no problem for the crocodile in its natural state, but in the contrived situation of having a human hold its mouth shut, the crocodile lacks sufficient strength to open its jaws against resistance. There are occasions when tackling a crocodile in this confronting manner may be considered legitimate (relocating animals away from areas of dense human population is a debatable business, but certainly far preferable to the old approach of simply shooting them). But you wouldn't want to lose your grip, because the next thing you would lose would almost certainly be your life (or a significant part of your anatomy!).

THE GAMUT OF EMOTIONS?

People sometimes misinterpret two facts of crocodile physiognomy, attributing human emotions where there are none. That's not to say that animals do not possess emotions — many species demonstrably do. Nor is it to say that animal emotions are discontinuous from human emotions; after all, we do share a common evolutionary history with the rest of the animal kingdom. Nonetheless, it's clear that the long, turned-up, toothy 'grin' of a crocodile's jawline cannot in any way be considered the equivalent of a smile. It does, however, allow us to differentiate between crocodiles and alligators, as the latter do not reveal their teeth when their mouths are closed.

What about tears, then? Most people are familiar with the phrase 'crocodile tears', which denotes a false show of feeling or empathy from a cold-hearted or manipulative human. This has garnered a rather more literary accounting in Edmund Spenser's *The Faerie Queen*:

> *[...] a cruell and craftie crocodile*
> *Which in false grief hyding his harmful guile*
> *Doth weep till sore, and sheddeth tender tears*

Underneath this literary edifice is a physiological basis. Crocodile eyes do tear up. Extrapolating from this the idea that crocodiles have a sentimental concern for their 'victim' (read: lunch) is in this case unjustifiable, but no doubt the sight of it would be compelling. Since antiquity travellers' tales have frequently mentioned weeping crocodiles, but the first written account was said to be in *The Voyage and Travail of Sir John Maundeville* in 1400: 'In that contre…ben gret plentee of Cokadrilles … Theise Serpentes slen men, and thei eten hem wepynge.' Spenser was writing in 1590, the height of the Renaissance, and would have been well versed in the various versions of this reported wonder, so appealing to the mediaeval mindset that preceded him. During his era, the phenomenon was variously interpreted as real or ersatz sorrow for the croc's intended victim, and the Spenser quote would be one of the earliest examples of its extension to metaphorical use applied to human beings.

On the whole, it is humans who travel. For the most part crocodiles are stay-at-home, stick-in-the-muds. Nobody knows the exact mechanism for it, but it is not uncommon to hear of relocated crocodiles finding their way back to home base. Equally inexplicably, the occasional adventurous estuarine crocodile ventures much further afield, even swimming hundreds of kilometres across the open ocean. But such an individual is rare.

Vox crocs

There are no handy decoders available for crocodile body language, such as those provided by Rick Aidan Martin (see page 105) for sharks. This is probably because their method of preying largely depends on the surprise element of sudden attacks. Prior to this, the most you are likely to have seen is seemingly immobile crocodiles basking on riverbanks or eyes shining by torchlight on the river at night.

Crocodiles do, however, have a fair range of vocal communications. In terms of reptilians, they are positively garrulous! A variety of differently pitched cries and growls can indicate threats, distress, contact, courtship — and, inevitably, hatching. One species can signal up to twenty different messages with subtle variations of these calls.

PROFESSOR SURVIVES THREE DEATH ROLLS

The late Val Plumwood was an internationally respected philosopher and activist. Her peaceful death at home in Australia in 2008 was widely reported in the international press, but not quite as widely as her close encounter with a crocodile in which she miraculously escaped death. She was one of the rarest types of philosophers — one who could prove beyond a doubt that her theories about humanity and nature are resilient, even after such an extreme experience of 'being prey' — the title of her account of this event.

Having been advised by a local ranger of the relative safety of being in a canoe off the main river system of the East Alligator River in Australia's north, Plumwood had spent a day or so paddling alone in the area's backwaters. In the afternoon of the second day, she saw a crocodile in the channel and made a split-second decision to take refuge in a paperbark tree on the bank. The crocodile also made a split-second decision and launched itself out of the water, biting and seizing its prey. During this ordeal, Plumwood was seized in the crocodile's jaws, escaped once, was seized again, and experienced the horror of three death rolls. Incredibly, she finally escaped and then managed to survive the long walk back to find help — a testament to her bushcraft and will to survive. In her own words, 'few of those who have experienced the crocodile's death roll have lived to describe it'. But she did survive and described her experiences with compelling honesty and economy.

One of the things that most dismayed her was the immediate response that someone should 'go out and shoot' the crocodile — a move she resisted fiercely. Professor Plumwood was a world-renowned scholar not a circus artist and her close encounter with the crocodile was neither staged nor exploited for its media potential. It was a serious — but to Plumwood, acceptable — risk attendant on her love of being in remote and wild places. The experience gave practical edge to her ecofeminist philosophy. When Plumwood writes about 'being prey' you know it is not just empty theorising, but the embodiment of lived experience.

WILDLIFE WARRIOR?

The late Steve Irwin does not qualify for a Damascus Award, mainly on the grounds that his heart — if not always his modus operandi — was in conservation right from the start. Despite the soubriquet 'The Crocodile Hunter' based on his television series of the same name, no one could accuse Irwin of having hunted wildlife with the intent to kill. He certainly hunted them for relocating, but generally his hunting was with a camera and sometimes over-exuberant enthusiasm.

Born in 1962, Stephen Robert Irwin was inducted early into a life of herpetology. His father taught him to wrestle crocodiles by age nine, and when he grew up he was one of the Queensland government's most successful crocodile relocators. He performed this service for free and was granted permission to relocate the 'problem' animals to the reptile and wildlife park founded by his parents and now famous as Australia Zoo.

Irwin was a showman, not a trained scientist, and he was neither sophisticated nor subtle in his persona. Apart from the infamous incident when he was filmed holding his baby son in one hand while feeding a crocodile with the other, Irwin was frequently criticised for exploiting and harassing wildlife. After his untimely demise some critics became a little more circumspect, perhaps not wishing to speak ill of the dead. Some, however, pulled no punches. The fearlessly iconoclastic Germaine Greer dared criticise the recently deceased. Both she and the hapless stingrays were mercilessly attacked after Irwin's death.

Love him or hate him, there is no doubt Steve Irwin was a dedicated animal lover and conservationist on his own terms. He certainly put his money where his mouth was, buying up tracts of land to preserve habitat and ploughing a great deal of money into conservation projects behind the scenes. A more unquantifiable aspect of Irwin's legacy is his impact on children. Many children's hearts and imaginations were captured by Steve Irwin, self-styled wildlife warrior. And some of those children may well be inspired to grow up to make great contributions to science and conservation.

COULD WE EVER SMILE AT A CROCODILIAN?

Apart from Steve Irwin, very few people are able to summon up much affectionate regard for reptiles, especially large and dangerous ones. The common antipathy to crocodilian species is understandable. Some members of the group are huge, ferocious predators whose modus operandi is the animal equivalent of a stealth missile. No fur, no feathers. Nothing soft at all. All in all, they would make for difficult neighbours. But there are equally good reasons to smile at a crocodile as at any other animal. Even they have an intrinsic right to exist, independent of human concerns, positive or negative.

Even from the human perspective, they are not without benefits. I'm not referring to the profitable exploitation of their hides which nearly drove them to extinction. There are ways, both tangible and intangible, that humans may benefit from their crocodilian co-inhabitants. These animals are diplomats from the remote age of dinosaurs, reminding us of the sheer wonder and diversity of evolution — and the possibility of living long and prospering if you successfully find a niche you can defend in dynamic balance. Of all the class Archosauria, only the dinosaurs vanished: the birds and the crocodilians live on. Maybe we could take note and become fascinated with what survived (and why) rather than what died out.

On a more practical level, their blood is worth bottling. Serum from wild alligators has proved effective against three human plagues: HIV, West Nile virus and herpes simplex. Alligators are efficient killers at the macro level, but the effectiveness of their serum against a very broad range of microbes is unparalleled. Again, these antiviral and antibiotic capacities may well be translatable into medical benefits for humans.

This notion of intrinsic wonder or even future benefit may be a little abstract compared with the deathly day-to-day risks of habitat sharing endured by African fishers since the upswing in the continent's population of Nile crocodiles from the late twentieth century on. As JoAnn McGregor, lecturer in Human

Geography, points out, you don't hear of too many Africans forming powerful lobby groups akin to those that resist the reintroduction of wolves and bears in parts of North America, for example. Yet the redefinition of the Nile crocodile from 'pest' species to animal in need of conservation poses potential dangers for the African communities that are far more immediate and real.

It's not surprising, then, that from some perspectives Western concepts of conservation look suspiciously like another form of neo-colonialism in which the locals get the shabby end of the bargain as usual. Certain approaches do resemble an updated, sugar-coated version of the Great White Hunter. Many people seem willing to countenance grave losses, providing they happen in the other camp. These are not simple issues, nor will they be easily resolved. It seems clear, however, that long-term conservation is only likely to enjoy enduring successes when it is seen as inextricably linked with, rather than in competition with, human rights and social justice.

UNDISPUTED KING OF DRAGONS

In the popular imagination, Komodo dragons have a lot in common with some of the crocodilians. They are large, primeval reptiles with the capacity to move fast, and a reputation as ferocious killers. In reality, though, crocodilians are only very distant relatives of the Komodo dragon (*Varanus komodoensis*). They share a taxonomic classification down to the sub-class,

Diapsida, but after that they followed different evolutionary paths. There is no doubt, however, that Komodo dragons and Australian estuarine crocodiles are two of the very few reptilian species that are capable of killing unwary or unlucky humans.

The legend 'Here be Dragons' on maps attests to the fine line between the known and the imagined. There are multiple theories about the origins of these mythical beasts that patrolled the borders in the world of the ancient mariners. The dragon that rules its island kingdom of Komodo, part of the Indonesian archipelago, might have been a contender, except that it wasn't known to the rest of the world until 1910. Locally known as oras, the Komodo dragon is also found on the larger Flores Island and a few other nearby islets. There are five small, separate populations in total, with relatively small reproduction rates.

The Komodo dragon is the world's largest living lizard: it can grow to 3 metres from its snout to the tip of its scaly tail, and a fully grown adult has an average weight of 100 kilograms or more. One spectacular individual reached 132 kilograms. That's big, which befits a lizard that eats people, but it pales in comparison with the now-extinct Australian monitor lizard (*Megalania prisca*) which is conservatively estimated to have weighed 600–620 kilograms. In their time these creatures were big enough to dine on pygmy elephants, as well as a range of other available prey. Although not mythical, this Pleistocene creature is long extinct, so that leaves the Komodo as undisputed giant of the lizard family and the only one to present a serious threat to people. And their bacteria are even worse than their bite.

HERE BE BACTERIA

Animals, humans included, that are bitten by Komodo dragons do not necessarily die immediately from the physical injury. Far more frequently the dragon leaves the injured party to brew up an infection or two and then returns later to finish the business. As the apex predator on an island, it has the luxury of leaving its meals to get into a desirable state of debilitation, or even downright rottenness. The dragon is well equipped in the department of distributing infectious diseases: one study identified 58 bacterial species in the saliva of wild and captive Komodos — and 54 of them were pathogenic. Four of the pathogens found in the wild dragons' saliva were resistant to antibiotics. An unfortunate Frenchman took two years to finally succumb to the infections he acquired from a dragon bite.

As if this wasn't bad enough, a recent research report published in the letters section of *Nature* added venom to virulence in the Komodo's repertoire of dangerous attributes. Australian researcher Bryan Fry and his colleagues discovered that Komodo dragons, iguanas and monitor lizards can all be added to the list of venomous species. Previously only the gila monster and bearded lizard (known as helodermatid lizards) were recognised as venomous. Fry's research posits a common clade (ancestral group) with a single origin for both the venomous snakes and the venomous lizards. Although Komodo dragon venom can kill smaller prey, it is unlikely to be the cause of death in humans, but it will inflict severe pain, breathing problems, skeletal muscle weakness and tachycardia.

THE ONES THAT GOT AWAY

Fishing for dangerous and deadly animals can throw up stories that sound like the classic fisher's fibs: you should have seen the one that got away; if only you were here yesterday you'd have seen some really big fish. Hmm. Only in this case it's true. And for 'yesterday' read at least 100 million years ago.

I confess that I was sufficiently surprised to discover, when researching this book, the sheer size of existing crocodile species, let alone prehistoric ones. As mentioned before, male estuarine crocodiles can reach 6 metres long and weigh up to 1 tonne. The largest Nile crocodile was alleged to be 9 metres, but realistic estimates put it equal to the estuarine species at 7 metres (a record length for the generally smaller Nile crocodile). This was, however, only the beginning.

Once I got going it wasn't so hard to picture the prehistoric 'terrible crocodile' — a 10-metre beast that inhabited the Columbian swamps during the Miocene period, around 5 to 25 million years ago. But it was the fearsome crocodile that defied my imagination. *Phobosuchus* measured 15 metres long and weighed 15 tonnes, with 10-centimetre long teeth. In *Phobosuchus, Tyrannosaurus rex* met its marine match. This is a comment on their similarly gargantuan proportions rather than speculation about any titanic battle of the beasts. Certainly, like tyrannosaurus, this Cretaceous crocodilian was large enough to eat some of the lesser dinosaurs with whom it shared an Age on Earth, although no one can say for certain that it did.

NEVER LOOK A GIFT DRAGON IN THE MOUTH

The world of diplomacy is a rarefied and strange one. Illegal poaching and trading in endangered wildlife is prohibited by the Convention on International Trade in Endangered Species (CITES). At the highest levels, though, breeding animals of such species are often gifted to other countries. There is no accepted traditional 'language' of animal gifts, as there is with flowers. Still, it does makes you wonder how much consideration Indonesia gave to its choice of a Komodo dragon for President George Bush Senior. Given the amount of bacteria in its saliva, it would never do to look a gift dragon in the mouth. The gifted animal was called Naga, a name which conjures up Hindu serpent deities. The president, perhaps feeling the White House lacked the necessary amenities for the 24-year-old lizard, donated it to the Cincinnati Zoo in 1990, where it was welcomed with open arms and provided with its own specially built enclosure. This particular Komodo dragon certainly had pulling power, and a ten-stop tour of other zoos garnered almost as much attention as a gubernatorial fiesta, gaining a lot of attention for conservation issues. Naga died of a stomach infection in 2007, but not before fathering 32 little baby dragons.

Naga's tour of duty was especially necessary as researchers have discovered that, left to their own devices, female Komodo dragons are given to parthenogenesis — asexual reproduction. Sounds reasonable, but it is of concern to zoologists. The problem with parthenogenesis is biological not zoological. This method of reproduction reduces the genetic variability of a population — and there is not a lot of 'play' in the system when it comes to Komodo dragons or any small population. There is a relatively stable population of 5000 animals wild in their natural habitat, as well as captive animals in various zoos and sanctuaries around the world. With such small numbers it's important to keep the genes mixing up on a regular basis to maximise their diversity.

ALLIGATORS IN THE MYTH

The dragon is one of China's pre-eminent symbols. Embroidered on silk, fired into porcelain, tattooed onto human flesh, the Chinese dragon symbolises power and good luck. But luck has run out for *Tu Long*, the earth or muddy dragon. In the year 2000 (ironically the Year of the Dragon in Chinese astrology) two US-based and five China-based researchers submitted a paper to *Biological Conservation* with the matter-of-factly grim title 'Wild Populations of the Chinese alligator approach extinction'.

Through the classically dispassionate scientific prose of the paper it's possible to see glimpses of an epic, but uneven, battle for survival, as throughout the twentieth century the Chinese alligator is forced back and back and back into more and more marginal habitat, until finally at the century's end there are less than 130 animals left — and most of those are isolated individuals rather than breeding groups. Both the estuarine crocodile and the Nile crocodile populations have made remarkable recoveries (too remarkable for some) — but their numbers never reached such a critical low. The one positive note is that Chinese researchers have created successful captive breeding programs, which means there is a source of animals to reintroduce into the wild. Most wildlife biologists realise that this is simply a stopgap measure, and captive-bred animals are necessary but insufficient for long-term conservation. The deciding factor is availability of habitat. Although the habitat of the Chinese alligator is legally protected, this does not always translate into real security of tenure. Active local protection of all remaining habitat is required. Without it, *Tu Long* will disappear into myth.

THE DAMASCUS AWARDS

Krystina Pawloski was an Australian schoolteacher with an unusual taste in after-hours activity. Although petite, Krystina was an excellent shot. She and her husband were both successful crocodile hunters. In 1958, she exceeded his achievements by shooting what is considered to be the largest estuarine crocodile ever seen. Apocryphally, the crocodile was 8.63 metres long and 4 metres around the middle. Although there is no indisputable evidence of this, it was clearly a large animal. Affectionately referred to as Krys, the dead crocodile is immortalised in a statue in the main street of Normanton, Queensland, where it attracts the usual run of tourists.

Far from trading on the potential notoriety of her killing skills, Pawloski instead turned her thoughts to conservation and quietly commented that, in retrospect, 'I was sorry the animal died for no reason'.

3

Venomous Virtue

Spiders, scorpions and ticks

Let's get one thing straight from the start: daddy-long-legs are not the world's most venomous arachnid. These inoffensive creatures have long been the subject of a pernicious urban myth that they would kill you if they could, lacking only a large enough bite to achieve the deed. Relax: daddy-long-legs are harmless to humans. This goes for both unrelated animals that go by the same common name: the true spider *Pholcus phalangioides*, and the 'harvestmen' from the Opiliones order. The latter are arachnids, but not spiders. (All spiders are arachnids, of course, but not all arachnids are spiders. The class also includes scorpions, ticks and mites.) Of course, if you insist on squashing daddy-long-legs then rubbing your eyes before washing your hands, you might get sore eyes. But why squash them in the first place? Listen to Issa, the seventeenth century Haiku master. He had the right approach:

don't worry spiders
I keep house
casually

It's a win-win situation — you get to be both ecologically sound and relaxed in your approach to housework at the same time. No wonder I used to feel smug knowing what I did about daddy-long-legs, but I'd barely scratched the surface. The tall-tale brigade has been far more inventive about these and other spiders than even I ever imagined: potted cacti that suddenly erupt and spew forth baby tarantulas (which, despite the old song, aren't even especially deadly anyway); baby spiderlings hatching in the hundreds after a spider-bitten arm swells up and bursts (people have seen too many *Alien* movies, perhaps?), and some of those beehive hairdos of the sixties didn't just harbour cockroaches or baby mice but spiders as well! All such stories are improvisations on the delicious–squeamish theme that predominates in playgrounds around the world.

Of course, it's pretty easy to pick the urban myths once you've heard a few. But even in the straight domain, the web of spider stories is very sticky and tangled. For example, I'm an Australian. I couldn't help but grow up believing that the 'deadly funnel-web spider' was, well, deadly. The deadliest spider in the world, as a matter of fact. No, no, no, say the Brazilians, we have the deadliest spiders in the world. The 'wandering spiders of Brazil', now they're a species to make an arachnophobe sweat. Well, reply the North Americans, our 'recluse spider' might be a bit shy, but one of those babies bites you and you're really in trouble. Then it's the red-back of Australia versus the black widow of many nations versus the katipo of New Zealand, until you discover they're all pretty much regional variations (different species) of the same genus — *Latrodectus*.

DEFENCE AGAINST THE DARK ARTS

Imagine attracting the title 'Doyen of Envenomation'. It does rather sound like a position vacant at Hogwarts School. But it is actually the honorific bestowed on Australian Struan Sutherland by James Tiballs. It refers to Professor Sutherland's stellar career as the researcher at the head of the Commonwealth Serum Laboratory (now CSL Ltd) where they developed the antivenom for the funnel-web spider and the venom detection kits that are now widespread in medical facilities in the developed world.

Just when you despair of ever finding out the world's deadliest spider you encounter your first true disbeliever and hear the argument that spiders don't really even rate when it comes to 'deadly and dangerous'. Don't dismiss this one before hearing it out. A number of experts are rightfully sceptical about exaggerating the danger of arachnids. The idea is not that spider bites have never caused a human death. Obviously, they have. It's more a suggestion that we look at the statistical rarity of such an event (zero deaths from the 'deadly' funnel-web since the introduction of antivenom in 1982, for instance). Then we look at how large spiders loom in our nightmares. Seen from this comparative perspective it does look just a wee bit disproportionate. Compared with snakes, which really do kill a substantial number of people per year, the spider doesn't rank as all that dangerous and deadly.

WHITE MISCHIEF?

If it's any consolation, being a person of science does not make you immune to the lure of a good spider story. Try uncovering the facts in the white-tailed spider (*Lampona* spp.) debate and you will find people weighing in with street stories and clinic stories, points of view from medicine and microbiology, zoology and toxicology. Certainly, some of these tales are disturbing. Quite a few people have documented evidence of extremely nasty skin damage — sometimes so extreme that it leads to amputation. It's just that very few, if any, of them have unequivocal evidence that the cause was a bite from a white-tailed spider.

Professor Julian White is one of Australia's pre-eminent toxinologists. He believes that Australia is responsible for the pernicious myth of the mischief caused by the white-tailed spider and dates it from 1982 when the possibility of necrosis (death of tissue) from white-tailed spider bites was put forward at the International Society on Toxinology World Congress in Brisbane. The only trouble was, as White points out, this was from a presumed spider bite. Good science, like good law, demands good evidence. But the white-tailed spider was not afforded the luxury of being presumed innocent until proven guilty. Like any other superstition, once guilt by association is well established, it's easy to work backwards from the clinical presence of skin damage to the unquestioned assumption it must be a spider bite, with little real evidence to back up the diagnosis.

Another of White's arguments is that this particular species was present in the population from the year dot, so why the suspicious absence of accounts of dire damage from white-tailed spider bites prior to the 1980s? Hmm. Come to think of it, like every Australian child I was brought up to be respectful and take proper precautions around red-backs (we don't have funnel-webs on the west coast where I live), but the dreaded white-tailed spider did not intrude on my consciousness until the stories of necrotising arachnidism began circulating in the 1980s. I must confess I succumbed to the myth. I've never bothered killing a red-back, but I've swiftly dispatched every white-tailed spider I found in my house in the last three decades.

But the era of death is over. I have seen the light in the form of an article published by Isbister and Gray in the *Medical Journal of Australia* 2003. Out of 130 confirmed white-tailed spider bites they studied, none produced the dreaded necrosis. Zero. Zip. Zilch. The bites might hurt a bit for a day or so. Some people may have a red lesion that lasts for a while. But no one has had an ulcer that lasted for more than a month from a confirmed white-tailed spider bite. If you want to avoid even this undramatic hassle then the best advice is shake out your laundry, look before you leap under the bedcovers and don't leave your clothes on the floor. All good scientific advice, which could equally well have come from your parents!

CONSIDER THE EVIDENCE

Still, that's not the same thing as claiming that all spiders are completely harmless to humans. There is no doubt that funnel-web spiders, as well as red-back spiders, black widows and the rest of their close relations, Brazilian wandering spiders and recluse spiders do pose a potential threat. However, they are also interesting creatures in and of themselves, and it's worth finding out a bit more about them.

Despite my complete commitment to reducing the amount of arachnophobia in the world, I have to say that the unofficial version of the US Army Survival Manual I came across on the Internet was just a tad dismissive when it said of funnel-web spiders: 'The local populace considers them deadly.' Quite. A little less than half an hour after being bitten by a funnel-web the majority of people are classifiable as severely envenomed. Victims will feel a tingling on their lips, twitchings of tongue and limb, as well as sweating, crying and salivating profusely — all due to the primary ingredient in funnel-web venom, robustoxin. The good news is that the funnel-web antivenom derived from *Atrax robustus*, the Sydney funnel-web, is effective against all the funnel-web species (and even in those rarer cases where people have reacted badly to the related *Missulena* spp. — mouse spiders).

Although the US Army Survival Manual advises avoiding funnel-webs, it only mentions a single genus — the *Atrax*. They don't know the half of it. The two genera of funnel-webs — *Atrax* and *Hadronyche* — contain about 35 species of spider. Funnel-web spiders are medium to large with dark-coloured bodies, large legs and large fangs. Both male and female build silk tunnels. Most, but not all of them, are ground-dwelling species. The funnel-webs are found all through south-eastern Australia, including Tasmania, but the majority of them have collected no indictments for human deaths. Not so the Sydney funnel-web (*Atrax robustus*) which is variously reported as having thirteen or fourteen deaths to its name since colonial times.

Not surprisingly, therefore, the Sydney funnel-web is generally put at the top of the most-wanted lists. However, recent research indicates that it is not this well-known species that has the highest capacity to inject venom. According to Isbister and Gray, the Sydney funnel-web only 'scores' severe envenomation

17 per cent of the time. A poor show compared with the southern tree funnel-web (*Hadronyche cerberea*), which achieves this effect 75 per cent of the time. The northern tree funnel-web (*H. formidabilis*) is no slouch either, at 63 per cent. On the other hand, there are a lot more people in Sydney and surrounds than there are living near the habitats of these latter species— and proximity is a big factor in risk assessment.

So by that measure does this mean that the recluse spiders (*Loxosceles* spp.) are not so risky? Or are they perhaps like the finally revealed serial killer about whom neighbours comment, 'He always seemed like such a nice, shy, retiring sort of guy, kept to himself mostly.' Also known by the nondescript common name brown spider and the more evocative violin spider, *Loxosceles* spp. often have a characteristic dark marking near their head that narrows down the back and appears to the more poetically minded observer to look like a violin. However, this marking is a bit hit and miss as an identifier, and it is much better to rely on the singular feature of all spiders in the group: the eyes. Most spider species have eight eyes to match their eight legs. *Loxosceles* spp. have six eyes arranged in a neat set of three pairs. You've heard the expression 'I've got eyes in the back of my head?' Well *Loxosceles* spp. have eyes watching the back of their abdomens — one dyad towards the front and one dyad each on either side. It might be better, though, to leave such close and personal observation to the scientists and naturalists who know what they are doing.

The average citizen is not very likely to get a chance to eyeball this type of spider, anyway. *Loxosceles* spp. aren't called recluse spiders for nothing. They are a nocturnal group found in several south-west and mid-west US states (thirteen species), as well as South America, Africa and parts of Europe. They like to live under bark or rocks, but have been known to find refuge in folds of cloth — casual housekeepers take note! It's no picnic being bitten by a *Loxosceles* spider. Their venom is neurotoxic (damaging to the nervous system), and at the very least will cause a nasty bite that is a long time in healing. Unlike with the white-tailed spider (see 'White mischief?' on page 70) there is clear evidence that the bites can be associated with necrosis, or death of tissue around the site of the bite. A small proportion of people bitten will suffer systemic envenomation with fever, chills, nausea, weakness and joint pain. However, you are highly unlikely to die unless you are already in a weakened condition due to extreme youth, age or immunosuppressed status.

The *Latrodectus* group, which includes red-backs, black widows and katipos, is generally considered to comprise at least 30 different species, although that number has shrunk and expanded like an accordion over the decades. Currently researchers are applying the techniques of molecular biology to sort out the tangled taxonomy. Whatever the final outcome on species numbers of this genus, it can be said that *Latrodectus* are truly spiders of the world. Like tourists, they are found just about everywhere and many *Latrodectus* species have taken advantage of human activities to expand their ranges. These hitchhiking spiders are not fussy about how they ride or where they end up. They'll be moved along with human belongings or cargo and simply take up residence, breed and spread in most new habitats. And like other sophisticated citizens of the world, they are given to some eye-popping sexual practices. The common name 'black widow' is not just a wild surmise. Female *Latrodectus* are generally considerably larger than males, and the male red-back stereotypically presents himself for consumption at the completion of mating, although other *Latrodectus* spp. may not do this — it's an area for future research. But from the perspective of a male red-back, at least, we are talking about a truly deadly spider.

What, then, of their impact on humans? According to the Australian Museum about 2000 people a year suffer bites from red-back spiders (*Latrodectus hasselti*), but nobody generally dies from them. In Australia nobody has died from a confirmed spider bite since 1979, and that was a singular event. In comparison, over nearly four decades in the United States there was a total of 63 deaths attributed to bites from the black widow (*Latrodectus mactans*). A red-back antivenom developed in Australia in 1956 is also useful for bites of the common black widow. Most of those bitten won't even require the antivenom but some will suffer considerably from the neurotoxic effect of the venom. Symptoms of *Latrodectus* envenomation include numbness, weakness, muscle pain, sweating, lymph gland swelling and hypertension. Not at all a pleasant experience, but with the advent of the antivenom, not likely to be deadly either.

The final group of spiders humans need to treat circumspectly are the Brazilian wanderers (*Phoneutria* spp.). They earned the name because of their habit of roaming about rather than staying in or close to their webs. They tend to be large (averaging 3 centimetres long, excluding their legs), and unlike most spiders they have the reputation of being aggressive, although hardly as aggressive as humans (see 'Horror Movie', below). *Phoneutria* venom is neurotoxic, and therefore quite similar to *Latrodectus* venom. Bites are extremely painful but usually not fatal.

So, that just about covers the spider species worth worrying about. Although the information may not cure severe cases of arachnophobia, it should at least convince you that spiders are not the ravening monsters of Hollywood horror movies. On the whole, spiders are no more dangerous than ticks. And not too many people have phobias or nightmares about ticks, do they?

HORROR MOVIE

One day a Brazilian wandering spider wandered a bit too far from home and wound up in the kitchen of a British pub, after hitching a transatlantic ride in a box of imported bananas. (Yes, yes, it's just like the old song. But let me remind you again 'Hides the deadly black tarantula' is poetic licence — tarantulas simply aren't deadly. Hairy and irritating, yes, but not fatal; not unless you're a dog, that is.) Brazilian wandering spiders, however, *have* been known to kill humans. And their alternative common name is the banana spider.

Now a British pub might not always be the friendliest environment for foreigners, let alone foreign arachnids. But the events that unfolded and were widely reported in the press qualify as a real horror story. It's true that the spider bit first, and that the bitten man, Mathew Stephens, showed extreme presence of mind by using his mobile phone to photograph the animal where it lay dead in the freezer. This was a calm and sensible precaution. Only the spider wasn't dead, just temporarily out of it. So, according to *The Times* newspaper, Mr Stephens went on to pour boiling water on the spider, collect it in a jar, and then — just to make sure — *microwave* it. The man was nothing if not thorough.

Even then, the spider didn't die. That would happen later, after hospital staff — showing more sangfroid, but less presence of mind — released the spider in the hospital gardens, not aware that it was a deadly import. Now that could have been the birth of another urban legend right there. But zoo experts put paid to it. So what did eventually overcome the almost indomitable spider? Sustained cold. It would not have survived long at all in the British outdoors.

All you arachnophobics to whom this sounds like the worst horror movie come true, pause, take a deep breath and think for a minute about how it looks from the spider's point of view.

TICK, TICK, TICK

If you are suggestible, please note that I am by no means recommending you should have a phobia about tick paralysis. It's just that ticks deserve about the same amount of caution as dangerous spiders. Ticks, along with mites and scorpions and spiders all belong to the same class, Arachnida, which is part of the phylum Arthropoda. Within each order of Arachnida there are species that can do damage to humans. But in terms of causing human deaths they are not even remotely in the same class as snakes! For example, according to *The World Spider Catalog* maintained by Norman I. Platnick of the American Museum of Natural History, within the order Araneae there are 40,024 species of spiders. Of these, only 25 are considered to pose any real risk to humans. A serious reality check is provided by an authoritative World Health Organization (WHO) report on an antivenom workshop. Written by some of the world's foremost experts, Theakston, Warrell and Griffiths, this report says that 'Stings by fish, cnidarians [jellyfish, hydra, sea anemones, and corals], lepidoptera [butterflies and moth], centipedes and cone shells and bites by spiders, ticks and one genus of octopus [that beautiful little blue Australian!], are responsible for some morbidity and mortality but probably account for not more than a total of about 100 deaths per year.'

TAKE NOTE

A straw poll of Australians would turn up very few who know the person whose image appears on their $50 note. Even when you tell them it is Sir Ian Clunies Ross only a handful of farmers and historians would be any the wiser. This is a pity, because this veterinarian was a major contributor to Australia's pastoral industry. And he is of more than passing interest here because he was also the person responsible for developing the antitoxin for the paralytic scrub tick (*Ixodes holocyclus*). This animal was a major plague for Queensland dogs in particular. In the tried-and-true method of developing immunity, he attached blood-swollen ticks to dogs and removed them after a short time. Eventually the dogs thus treated built up a resistance. The resulting serum was then able to be used as a treatment for those suffering from tick paralysis. This antitoxin was released in 1938 — only eight years after the development of Australia's first antivenom against the tiger snake in 1930.

Groundwork for the latter had been developed by Dr Frank Tidswell and Professor Charles Martin in the early 1900s. Tidswell also did pioneering work in tick research, as well as the venom of red-backs. It wasn't until 1956, however, that Australia provided the world with an antivenom for the red-back (and hence black widow species everywhere), courtesy of Saul Weiner of the Commonwealth Serum Laboratories.

Tick species — all 840 of them — are much smaller than their other arachnid cousins. But we certainly shouldn't be complacent about their deadly potential. The oral secretions of the tick itself can cause paralysis, and they also carry some pretty pernicious infective agents that cause diseases in humans: Lyme disease; Rocky Mountain spotted fever; human anaplasmosis; babesiosis; human monocytic ehrlichiosis; southern tick-associated rash illness (STARI); Q fever; and tick-borne encephalitis to name some of the better-known ones. Ticks are second only to mosquitoes in terms of their success as spreaders of disease. And courtesy of climate change they are on the move. Disease-bearing ticks are moving upwards and onwards into areas that were previously too cold to support them, including Sweden and the mountainous regions of the Czech Republic.

In the United States, it is deer that are the most likely involuntary source of disease-carrying ticks. But ticks aren't picky — be they deer ticks or bear ticks (which are actually one and the same, *Ixodes scapularis*) or any other sort of ticks, they will happily species-hop from deer to dog to you via any lizard that's lounging in between. Mammal, bird, reptile: ticks are equal-opportunity feeders. Just bring on the blood. And preferably by the bucket load: although tiny (sizes range from 3–23 millimetres for various species of ticks, including *Ixodes*, *Rhipicephalus*, *Amblyomma* spp.) they are no slouches when it comes to feeding. Some ticks have been known to swell to more than twice their original body size after feeding. Rather than similarly inflate your fear, it's worth remembering that most tick bites do not result in tick-borne illness, and that vigilance reduces the risk even further.

TERRORIST TICKS?

What is more of a biological time bomb than random tick bite, according to tick expert Professor Stephen Wikel of the University of Connecticut's health centre, is what he considers to be the very real threat of bioterrorism by introducing serious pathogens into tick species. The longevity of ticks, their capacity to travel courtesy of their hosts, and the ability of some species to pass pathogens not just to their host but also to their offspring, makes this a very dangerous scenario. It's behaviour like this that makes *Homo sapiens* contenders in the deadly animal stakes.

STING IN THE TAIL

The WHO researchers quoted on ticks above are a little less sanguine about scorpions, which in their esteemed estimation rank as second only to snakes in terms of their danger to humans. Even so, snakes cause around 80 per cent of all envenomation deaths, and scorpions account for 15 per cent so they are a pretty poor also-ran in the mortality stakes. The total number of scorpion species is frequently given as 1260, whereas accounts of the total number of snake species worldwide range between 2500 and 3000. At least half of the scorpion species are venomous but only 50 or so are considered a serious medical risk. All but one of the venomous species belongs to the family Buthidae (the exception is the *Hemiscorpius*, family Hemiscorpiidae). Not for nothing is this group referred to as 'thick-tailed' scorpions. They have relatively feeble pincers but their stinger is built for business. The other three scorpion families are vice versa, with thick pincers and thin stingers as they prefer to crush rather than paralyse their prey. This only presents a risk if you are pretty small. This fact is no doubt known to one Ali Khan of Malaysia.

THE SCORPION KING OF MALAYSIA

Khan, the self-styled 'Scorpion King' and 'Cobra King' of Malaysia, once spent 21 days in a glass cage with these allegedly 'deadly' animals. Added to this were a few thousand extras from the wild, dropped into the cage by a sceptic who thought the original animals were way too tame. Although the resulting video clip of scorpions crawling all over Khan makes disturbing viewing — from more than one perspective — it's quite clear that he was canny enough to cohabit with thick-pincered species. Although he emerged with a number of stings earned in his bid for fame, his life was hardly endangered by the experience.

Even the thin-pincered Buthidae take a minor toll on humans, compared with snakes — less than 1000 deaths have been noted worldwide out of around 100,000 incidences of scorpionism officially recorded. Of course, as with snakebite, this would underestimate the total figures, with reporting systems being far from comprehensive, especially in developing nations, but it still shows the comparatively mild risk of dying from a scorpion sting. It's just as well, because although scorpions are predominantly associated with deserts and dry arid regions they can be found from the wet tropics to the icy Himalayas and everywhere in between. And, just as with the Brazilian wandering spider, some have been known to hitch a ride in cargo ships or luggage.

In fact the travelling habits of scorpions have recently embraced domestic airlines in the United States, where in a single week in January 2007 there were two separate instances of mid-air scorpion stings. The first was relatively explicable as the victim had been camping in Costa Rica, (where the undeclared, undetected animal must have crept into his haversack), before flying home to Toronto. So much for enhanced airport security! The other case was a bit more mystifying, and no theory has really been advanced as to how the scorpion got to be on the flight from Chicago to Vermont. It would be easy to imagine that the unfortunate passenger was then at equal risk of developing a phobia about flying or a phobia about scorpions, or both. By all accounts, however, he seemed pretty phlegmatic about the event. You just can't rattle some people.

Although the occasional jet-travelling scorpion attracts more media attention, it's a fair bet that you run a greater risk of scorpion envenomation at home by native species if you live in Mexico, east-central South America, North Africa, the Middle East or India. And it is also likely that these regions have far higher numbers of cases that go unreported. There's some good news, though. A major WHO campaign launched in 2006 will help make antivenom much more widely available in those regions most in need. Although the jury is still hung on the value of antivenom as the best line of defence in scorpionism, it may have a role to play, especially when envenomation is severe; and it is certainly used in some parts of the world. In Turkey, deaths from scorpion envenomation have fallen dramatically since new protocols and supplies of antivenom were made available. This is no doubt cause for celebration in Birecik, a village that seemed under siege by scorpions, having suffered close to half of all scorpionism cases in the region.

On the other hand, a strong contingent of medical researchers, particularly from Israel and the Indian sub-continent, argue that antivenom is overused and inappropriate in many cases, mostly because scorpion venom takes effect much more swiftly than that of snakes, so it is usually a medical case of shutting the stable door once the horse has already bolted. Another tack is the possibility of prophylaxis, or preventative measures. The Mexicans are keen on this option, but it is yet to be developed in any scientific way.

MYTHS AND SYMBOLS

Like their venomous brethren snakes, spiders and scorpions seem to attract about an equal amount of ambivalence when mythologised. Good or bad, they are all liminal — creatures of the threshold. The whole biological classification Arachnida is named for the Greek myth of Arachne. Arachne was a Greek girl gifted with superlative weaving abilities and little political sense. She challenged the goddess Athene to a contest of skills and won. Needless to say this didn't go down well with Athene, so the goddess tore up the winning tapestry and turned Arachne into a spider for her troubles.

The West African trickster figure Anansi is a spider god. Tales of Anansi have travelled far and taken root in various cultures around the world. In some of the southern parts of the

United States spiders are referred to as 'Aunt Nancys', which isn't too far from Anansi, whose stories were told and retold by people of African origin forcibly transported to the region during the era of slavery.

There are also many strange tales of salamanders and scorpions. Salamanders and phoenix are supposed to thrive in fire, but there is an ancient legend that claims scorpions will sting themselves to death when confronted with flames. It's not a myth that holds up under scientific scrutiny, because suicide by their own venom is not an option — its own venom will not kill a scorpion. The more likely explanation is that the animal is being tormented by the heat from the flames and in its writhings appears to be stinging itself — a truly hellish vision!

There is a lot of hype in the common names given to scorpions: southern man killer, death stalker and — my personal favourite — the giant, desert, hairy scorpion. Typically, however, the only really dangerously venomous scorpion species found in the south-western parts of the United States is called by the innocuous common name 'bark scorpion'. This Centruroides species can kill but doesn't do so very often. From an annual average of around 13,000 stings, the United States only records one death from scorpion envenomation every two or three years. This is almost always a small child. South Africa, too, has a relatively low death toll (less than half a dozen annually; again the victims are mostly children). Various Parabuthus species are considered responsible for these deaths. Although there are other equally — if not more — venomous scorpions in South Africa, Parabuthus are more dangerous because of their preying habits. Unlike other species that are opportunistic and passive predators, Parabuthus actively forage, which naturally increases the chances of a human–scorpion encounter.

It's difficult to imagine, but these hard-looking creatures actually have a reputation for being good mothers, too. In contrast to most arachnids, scorpions are viviparous — they produce live young, a significant number in some cases. *Androctonus australis Hector*, for instance, produces up to 90 scorplings per brood. These scorplings will live on their mother's back and be dependent on her until their first moult five or six days later. The common striped scorpion (*Centruroides vittatus*) carries a pregnancy for almost as long as humans — eight months to be exact. And two researchers from the University of Texas, Lawrence Shaffer and Daniel Formanowicz, Jr., found that just like heavily pregnant humans, pregnant common striped scorpions cannot run easily, but they will stand and fight if necessary in defence of themselves and presumably their unborn young!

BLACK LIGHT

Night-time is the right time for scorpions. They are most often nocturnal hunters that prefer to hole up in any given nook or cranny during daylight hours. Given these habits, pretty much the same precautions that apply to avoiding snakes and spiders also apply to avoiding scorpions. Predictably, most scorpion stings are on the extremities. Wear shoes, particularly outside at night; understand that what you consider to be your woodpile, other species consider to be their home; refrain from putting your hands or feet anywhere you can't see clearly. These are all helpful suggestions for avoiding venomous species. Oh, and there is one final tip that is exclusive to scorpion avoidance: one of the flashier habits of scorpions is that they fluoresce under black light. So if all else fails, you could light up your garden like a seventies disco to make sure you spot the scorpions before they accidentally sting you.

Just as snakes shed their skins, scorpions need to moult their exoskeleton, a process that happens a minimum of five times before a scorpling reaches adulthood. From the time the old exoskeleton splits to the time the new one hardens, the growing animal practises a form of scorpion yoga — extending their flexibility to the limit so they won't be too constrained by their new 'suit of armour'. In the transition phase before the exoskeleton hardens they are at much greater risk of predation. Only when the process is complete will their fluorescence return. We could take a tip from the scorpions and work at not being too trapped by our rigid preconceptions about dangerous animals!

GREEN LIGHT

In all circumstances it's really good to resist succumbing to SLOPS (severe loss of perspective syndrome). This is never more relevant than when thinking about the animals that we like to be scared by, in this instance the arachnids. Treatment is available for people whose arachnophobia is so severe it is ruining their social life. As for the rest of us, a few simple precautions are enough to reduce the small threat some species of arachnids present. And — as arachnophiles can tell you — the vast majority of arachnid species are far more threatened by humans than vice versa. Spiders are a perfect example: there are 40,024 species and less than 0.6 per cent of those species have the capacity to envenom humans. The other 99.4 per cent provides us with immense benefits by consuming insects that cause untold human misery in terms of disease and damage to crops and food stores. Around the world loss of habitat means the loss of diverse species assemblages, some of them as yet unknown to science. It might be worth turning the green light of conservation onto this fascinating group of our cohabitants on planet Earth — they don't take up nearly as much room as we do.

4

Outrageous Fortune

Sharks and stingrays

In his lifetime, a certain Swedish gentleman was famous for taking the Chinese invention dynamite and applying it to various practical applications, including military ones. When a newspaper mistook his brother's death as his own, Alfred had the rare and peculiar experience of reading his own obituary.

Horrified at the prospect of going down in history as a merchant of death, he promptly set up the Nobel Peace Prize, which now brings honour to his name. On a much smaller scale, the Benchley Awards for excellence in shark research do much to counteract the hysteria and ignorance fanned by both the book and the film *Jaws*, written by the same Peter Benchley. Human terror of sharks pre-dates those works, but the book and film certainly capitalised on the existing fear with such exquisite exactitude that the very word 'jaws' is now the cultural yardstick, a shorthand term that invokes shark-induced terror. Oddly enough, Benchley has become a lifelong fan of sharks and argues strongly for their conservation, hence the research awards. More research is sorely needed.

While we have pretty accurate figures for human populations, and a rich knowledge of human demographics, cultures and dwelling densities, the same cannot be said of sharks. The oceans and their denizens remain vastly mysterious and enticing. No one agrees about the total number, but scientists have described more than 450 species of shark alive in the world today. Fourteen of these are known only from a single specimen or extremely rare sightings. The extraordinary megamouth is one such animal. First discovered by humans in 1976, there have been only 37 confirmed sightings subsequently. Almost certainly more shark species are yet to be discovered. What we do know for sure is that at least 80 species of shark are considered endangered, and a greater number are classified as threatened species. This list includes the great white, which occupies the unenviable position of being in the top ten species most compromised by international trade, according to the World Wildlife Fund (WWF), as well as appearing on the International Union for the Conservation of Nature and Natural Resources (IUCN) Red List as a 'vulnerable species'. The IUCN is systematically evaluating the status of all sharks and rays. Twenty per cent of the species reviewed so far are considered to be in danger of extinction. There are some who might find it difficult to mourn the loss of sharks, but until fairly recently, the ray family was less inclined to receive bad press.

STINGRAYS

Certain species of rays lie hidden in the mud and silt of both ocean and sweet water homes and some undulate like dream angels through the upper reaches of the open ocean. As a girl snorkelling alone close to the shoreline I saw a ray materialise as if spontaneously generated from the seabed below me. Oblivious to any danger, I swam behind it for a few metres. Across the intervening years I am still hypnotised by the image of the slow rise and fall of its dark wings. It was a singular experience for me, but my ocean-going brother speaks of the reverse. In his spear-fishing days he was frequently followed by rays that appeared to be both friendly and curious. Not to mention hungry, and no doubt happy when tossed the occasional fish from his catch.

Times and values have shifted. Current expertise advises against humans initiating contact with wild animals. Still, of those who spent so much time underwater during the heyday of spear-fishing in the 1950s and '60s, some experienced and learnt a lot about the habits of the creatures whose realm they entered. It was common knowledge to them that stingrays were armed against their prey, and that dark shapes swimming above a ray would almost certainly evoke a response designed to disable the shark, which is both its closest relative and most feared predator. It was the received wisdom that the barbed sting could cause severe pain — and even death if it penetrated the chest cavity. Despite this, most agreed that given a good mix of commonsense, caution and respect, temporary co-existence with marine animals was both wondrous and relatively safe.

OUTRAGEOUS FORTUNE

The vast majority of stingrays are harmless but some, given very specific conditions such as swimming directly above them at close range, can occasionally be fatal to humans. Unfortunately for them, the retiring stingray has gained unsought notoriety after the death of celebrity Australian conservationist Steve Irwin in 2007. Vicious and senseless stingray massacres followed his death: hardly a response that Irwin himself would have supported. The stingray that caused his death is likely to have been a member of the Dasyatidae species, possibly *Dasyatis brevicaudata*, one of several species that go by the common name of bull ray. These animals are certainly capable of delivering a severe sting, but it is only when that sting happens to penetrate a major organ (in Irwin's case, the heart) that death follows. Even then, there are records of people having survived a stingray injury to the heart.

Lesser injuries from accidental encounters with stingrays are far more common. The barbed spine or spines of the animal's tail can cause deep and ragged-edged wounds that are difficult to heal, especially when bits of the spine have broken off and become embedded in the flesh. Pulling the sting out is likely to cause worse injuries than the entry wound. If the person is lucky, the stingray may be unable to inject venom because many of them lose the thin layer of skin that covers their venom glands, thus rendering the glands useless. When venom is injected, the pain can be unbearable. There is also the possibility of a variety of opportunistic infections. The venom of some stingray species has been recorded as causing necrosis (dying off) of human flesh. One exceedingly unlucky person actually survived an original stingray injury to his chest cavity, only to die suddenly some time later when the resulting necrotic infection eventually caused cardiac failure.

These dramatic and dreadful incidents are exceedingly rare, and should not be given disproportionate emphasis. Short of the shark phobics' radical option of swimming only in pools built by humans, a few simple precautions can dramatically reduce your chances of any unpleasant encounters. Remember that these animals are basically non-aggressive, so give them a chance to get away from you by shuffling your feet along the bottom when walking in shallow areas. As on land, putting your hands or feet somewhere your eyes can't see is asking for trouble. Finally, since Steve Irwin's untimely demise it ought to be clear to everyone that people should avoid swimming above stingrays.

That does not preclude swimming *near* rays. Many marine scientists have spent pleasurable and fruitful hours doing just that. And their observations have taught us a lot about the behaviour of these compelling and intelligent animals to add to our existing knowledge of their evolution and morphology. Rays are Ikea sharks — they come in a flat pack. Although not all researchers agree on the details of the various complexities of the relationship between the two groups, it seems clear that rays and their relatives are simply another branch of the elasmobranchs. Going by the collective name of Batoidea, the earliest members of this younger group first made their appearance on Earth 200 million years after the sharks. Batoidea comprise at least 500 or so extraordinary and beautiful member species — more than half the world's living elasmobranchs. Electric rays, sawfishes, skates, stingrays and guitarfish are all members of the group.

FROM DISK TO DIAMOND

Batoid 'wings' are actually attenuated pectoral fins; as if to counterbalance this, most batoid species have dispensed with tail and dorsal fins altogether, although a few possess rudimentary ones. Guitarfish are the morphological halfway house: the distinctive shape that earned them their name is flat and ray-like at the head — and stout and shark-like at the tail. As the species evolution radiated the batoids developed a variety of body shapes from disk to diamond.

Human parents and teachers may claim to have 'eyes in the back of their heads', but the batoids literally have eyes on the *top* of their heads. They also have spiracles (openings through which water and air pass as the animal breathes) on the top of their body, behind the eyes, which makes breathing feasible when they are lying in the mud. Their eyes give them a good 180-degree view of the water above and the ground around, but rays and skates are blind to what is below. However, that doesn't mean they are unaware of what is going on underneath them. Alternative sense mechanisms — both olfactory and electrical — mean that they have a very good idea of what prey is doing below (imagine them admonishing their startled offspring with 'Don't even think about it, I have a nose on the bottom of my body!' Actually, they don't have noses as such: their odour-detecting system is located just above their mouth.)

As well as sensing prey, the electric sense system of some species is also used for social interaction. Like prospectors who use metal detectors to search for buried treasure, some rays use this extra sense to seek buried mates. Strangely, despite their flattened bodies, rays have bigger brains than sharks, and many Batoidea species are highly social animals. Some species congregate in large social gatherings, others journey together in huge schools.

PREY TELL

Although it it fairly rare for rays to kill people, almost any shark species has at least the potential to inadvertently harm humans, given the right circumstances. Even the whale shark — that gentle vegetarian — has been known to knock unwary humans unconscious. Still, if foolhardy or fascinated divers are so keen to get up close and personal that they fail to watch out for the animal's massive dorsal fin this hardly constitutes malevolence on the part of the shark. In fact, malevolence doesn't enter into the equation. Predation is a fact of life, and sharks are superbly adapted predators. Humans don't even rank as shark prey. Shark attacks on humans are attributable to other causes: mistaken identity, competition for food (official records reveal that around two-thirds of shark attack victims were spear-fishing or carrying fish at the time of their attack); and accidentally or deliberately overstepping the boundary of the shark's personal territory.

Forty-two species of shark have been documented as attacking humans; several species are considered particularly dangerous, including makos, blue sharks and oceanic whitetips. Only three shark species, however, are serious contenders when it comes to tallying up human fatality statistics. Many of their common names reflect this:

WHITE SHARKS, also known as great white, white pointer, and white death (*Carcharodon carcharias*)
TIGER SHARKS (*Galeocerdo cuvier*)
BULL SHARKS (*Carcharinus leucas*).

CRUNCHING THE NUMBERS

Worldwide statistics are kept by the International Shark Attack Register, now hosted at the University of Florida. Florida is an apt home for such a database, because it is shark attack central for the United States of America. (Mind you, even in Florida you are six times more likely to be struck by lightning than you are to be bitten by a shark.) The register was started in 1958 partially in response to documented shark attacks on US navy personnel during World War II. Prior to this, scientists considered sharks to pose no threat to humans unless they were injured (the human that is, not the shark). The keepers of the register were nothing if not thorough, and the project has retrospectively collected historical data as far back as the mid 1500s. The database now includes details of more than 3000 shark attacks. In terms of current data, there is the possibility of under-reporting — both from undeveloped regions of the world, as well as highly developed tourist destinations that wish to avoid bad publicity. All in all, though, the Shark Attack Register statistics are considered robust. They also reliably demonstrate changing trends from year to year.

We know from the register that the average number of human deaths from shark attacks worldwide is between 50 and 60 a year. Two factors that might have shifted the balance of this average over the last half a century have, in fact, worked on either side of the equation to keep the figure relatively stable. First of all, there is a trend for shark attack fatalities to increase just because of increasing human population, and shifting patterns of human recreation. It is simply a numbers game. The more frequently people spend time in the ocean, and the longer the time they spend there, the greater the chances of encountering the sharks whose home environment is being trespassed upon. On the other hand, denser human populations — and the infrastructure that accompanies them — means a greater chance of surviving a shark attack.

The whole idea of a shark attack induces so much terror that many people can't imagine anything beyond a single, deadly bite. They would be amazed to know that among the already vanishingly small number of people who are attacked by sharks, even fewer actually die. Most survive the encounter, albeit some with severe damage and missing limbs. Unless you're in Australia, less than one quarter of shark attacks has a fatal outcome for the humans (472 out of 2119 to be precise, according to the International Shark File Register in 2007).

The United States and Australia, closely followed by South Africa, top the charts when it comes to shark attacks, with one crucial distinction. For some reason, slightly more than one-third of shark attacks in Australian waters are fatal. The difference between life and death is usually access to immediate and informed first aid, and swift transport to excellent emergency services and medical help, so perhaps it is the remoteness of some of the Australian surfing haunts (Cactus Beach, on the Great Australian Bight, for example) that contributes to the higher death rate. In general, though, as the figures on one side of the equation are going up — more people in the water, more shark encounters — improved emergency procedures weigh in on the other side and keep the fatality figures well below a hundred a year, worldwide.

Clearly, even now, the likelihood of suffering a fatal shark attack is extraordinarily remote. And it can be eliminated entirely — as many shark phobics will attest — by simply staying out of the water!

Even if you don't take the abstinence option, the chances of being attacked by a shark are practically nil. But the consequence if you are is extremely serious. (The people who study risk management have a term for it: zero infinity.) The loss of any human life is an infinitely incalculable tragedy for those left behind. This makes it even more amazing that the relatives of two men killed by sharks in Australian waters in the first decade of the new century — Brad Smith of Western Australia and Nick Peterson of South Australia — were somehow able to keep a broader perspective even in the midst of their immediate loss and grief. Both Smith's brother and Peterson's father made public statements resisting the call to hunt or exterminate sharks. In doing so they affirmed their family member's love of the ocean, and willingness to take it on its own terms. They also swam against the tide of public opinion, which is frequently fear-driven and unable to see past vigilante notions of revenge. Such perspective is to be admired, especially under the difficult circumstances of human death. According to Peter Benchley, for every human death caused by a shark, 10 million sharks are killed by humans. This means that the sum total of sharks of all species killed purposely or accidentally every year must number in the hundreds of millions.

THE JOURNALIST'S MEASURE

For every person bitten by a shark each year, 25 people are bitten by New Yorkers. (It is not recorded how many sharks have been bitten by New Yorkers.) But the original Great White Way — the Hudson River — is no longer home to the great whites that lived there in the 1920s. So, alas, they have been bitten fatally by New York, if not by members of its populace.

SPEAKING SHARK FOR BEGINNERS

As a young girl I hero-worshipped my elder brother, a dedicated spear-fisherman. He taught me to snorkel and dive, thus giving me the keys to the wonderland under the ocean. Long before ecotourism was heard of, I swam with Australian sea lions and dolphins and rays (which I now know is not recommended if you are keen to avoid sharks). I was conscious of sharks, but not paranoid about them. I treasured the one piece of shark knowledge I had: if bitten by a wobbegong, there is a moment when it gives a characteristic 'cough', and this is your only chance of removing your limb from its bite. Age brings humility. And now I'm not so sure I would have the presence of mind to make the most of this knowledge during an actual attack. Still, it's probably just as well to know the theory of how to speak shark. With luck, very few of us will be called upon to demonstrate a practical understanding during a direct encounter.

There is a consensus among shark experts that there are three basic styles of shark–human attack, each of which signals something different. By far the most common is probably the result of mistaken identity on the part of the shark. Known as the 'hit-and-run', this kind occurs most often in shallow water. The shark may bite and let go, or slash an arm or a leg, before disappearing. The perpetrator of a hit-and-run is unlikely to persist, having discovered that what it has bitten is not really to its taste. The person attacked seldom sees the shark coming, and may not register immediately that they have been injured. Scientists suspect that most hit-and-run attacks are made by young sharks that have not yet finessed their ability to accurately identify prey species; some may even be the equivalent of a temper tantrum. If you thought human toddlers going through the terrible twos were a handful, you don't want to know about the shark equivalent (an emotionally out-of-control infant might embarrass, exhaust or even enrage you, but an infant shark of the right species has the power to kill you without even meaning to). On the other hand, the hit-and-run is the shark attack you are most likely to survive.

People in deep water can also be taken by surprise, and for that reason the second style is known as a 'sneak attack'. Unfortunately, in this instance the first bite is not the last, and the shark will persist. It may prefer a meal of other species, but it will be large enough to eat a person and willing to take what is on offer. The same is true of the final style of attack, which also occurs in deep water. Only this time, there is a warning. The 'bump and bite' attack may — as the name suggests — be horrifying, but it is not without advanced notice. In this category the shark will circle and bump its intended prey before biting. This does at least provide some opportunity to deploy any and all of the survival tactics you can summon up. These can work, as Tim Dicus attests. Dicus, a surfer who went to the rescue of a fourteen-year-old girl who was attacked by a shark off Miramar Beach in Florida, said: 'You always hear, "Pound him on the nose." Then you look at an 8-foot shark and think, "What?" But it worked.'

Sadly, despite Dicus' heroic rescue attempts, the young girl did not survive.

There was a happier outcome for Jason Cull, a 37-year-old schoolteacher from Albany, Western Australia. Cull was attacked by a 4-metre white pointer while swimming with dolphins off Middleton Beach in May 2008. Thinking his attacker was a dolphin, Cull reached back and attempted to drive it off by poking its gills. He missed and hit the shark directly in the eye. This was enough to make it let go and Cull headed for shore in imminent danger of being re-taken by the same shark or one of its two companions. Showing enormous courage, volunteer lifesaver Joanne Lucas swam out and assisted the badly injured man. Cull recovered after surgery, and Albany's beaches were re-opened after eleven days when the sharks were no longer in the area. The moral of these stories? If you are ever in the dreadful position of being attacked by a shark, you have nothing to lose by poking or hitting its vulnerable spots (nose, gills, eyes) as hard as you possibly can.

THE CLEAN BOYS

When I was fifteen I topped off several years of swimming lessons by going for my Bronze Medallion in lifesaving. I was one of only two girls in a group of two dozen candidates. I was not an especially strong swimmer, unlike the tanned and hulking youths who powered up and down the pool, overtaking me easily. But I did possess one advantage: I listened. Every day for two weeks the instructor would patiently drill us in lifesaving theory, part of which involved answering the question: If someone is bleeding from a shark attack and all you have to stem the haemorrhage are dirty old rags from your car boot, what should you do? For some reason my brawny classmates, who likely spent their entire weekends tinkering with dirty car engines, suddenly turned fastidious. Under no circumstances could they be brought to see that allowing somebody to bleed to death because of being squeamish about a bit of dirt was not going to win them any lifesaving awards. Antibiotics can do a lot, but they are no help to a 'clean' corpse.

SPEAKING SHARK FOR THE TRULY COMMITTED

The dedicated few who have observed sharks closely over many years are able to comprehend a range of subtle shark communication modes far more comprehensive than the three simple patterns outlined above. One such person was the late Rick Aidan Martin. Martin originally hailed from Queensland, Australia, where snakes and sharks could be considered part of the normal hazards of an adventurous childhood. He began his experiments in natural history early in life and carried on with them to great effect until his untimely death at home in Vancouver, Canada, in early 2006. Given that Martin developed his comprehensive vocabulary of shark signals by dint of working with another diver (each of them would take it in turns to swim at a shark, leaving it with no escape route!) it also underlies just how exceedingly unlikely it is to be killed by a shark, even if you spend your professional life pursuing them. Martin left a great legacy from a lifetime dedicated to research and conservation. Divers, especially, are grateful for his posthumously published paper that describes 29 different elements of shark threat displays, some of which are widespread among different sharks, and some of which are species-specific. These range from the more widely known (at least among shark experts) to information that has never before been observed or scientifically recorded.

Most threatened sharks will demonstrate their distress or dislike by adopting a position known as 'the hunch', a bit like a cat arching its back. The pectoral fins are oriented down, but the head will be up; the shark is giving you the 'heads up' that it is not happy with your presence or your behaviour. This position is hard to miss in some species, such as the grey reef shark, but blink and you will miss it in the great white which might only hunch for three or four seconds.

IF IT AIN'T BROKE, DON'T FIX IT

Sharks perfected the basics of their biology early in evolution, and having settled on a winning combination they've pretty much stuck with it for the last 400 million years or so that they have been on Earth. They share a great deal in common with all the other fish, but together with the rays and the chimaeras, they are classified as elasmobranchs — cartilaginous fish, in common parlance. The elasmobranchs are a tiny group (representing about 1 per cent of all fish life). Here are a few distinguishing features of their biology:

Sharks do not require a swim bladder, as their buoyancy is derived from an oil-filled liver. The combination of light, flexible cartilage with the absence of a swim bladder enables sharks to move up and down the water column at a rate impossible for the bony fish. It's as if they have access to some super-fast elevator.

Although they have no vocal cords, sharks have a wide range of communication tools at their disposal. The body language described in 'Speaking shark for beginners' on page 101 is extraordinarily detailed and both species- and situation-specific.

Both hearing and eyesight are excellent, and they have senses unavailable to humans, particularly the ability to decode electrical messages through an organ known by the mediaeval-sounding name ampullae of Lorenzini.

There's also something special about shark jaws. Like the snakes that can unhinge their lower jaws and thus swallow disproportionately large prey, sharks can protrude their upper jaw during feeding. Although the shared capacities are derived from entirely different evolutions and mechanisms, the widely different species both derive a similar benefit from the moveable jaw.

Beating the rap

Other signs of shark displeasure potentially preceding an attack sound like a catalogue of hip-hop dance moves. Martin sorted out the anecdotes from the reliable scientific evidence and came up with a definitive, although not exhaustive, list of 29 different possibilities. These include tail slapping, tail popping, charging, rolling, tilting and shivering, as well as the aptly named RAP (repeated aerial gaping). In fact, the analogy is not without a deeper significance. All such behaviours are termed agonistic by biologists. In other words, it is that competitive, ritual signalling behaviour that many animal species use to threaten and ward off challengers. Just like any group of homeboys, rapping and popping to play out street rivalries. And just like on the street, it can be either for show, or in deadly earnest. And certainly Martin's article is prefaced with a significant warning that boils down to the fact that a little knowledge is a dangerous thing, and knowing something of sharks' behaviour is no guarantee of protection against them. I'm glad I never had to test out my little scrap of wobbegong folklore in reality. I'm sure I would have been the one to end up woebegone! Far better to avoid the first bite.

TABLE MANNERS FOR TIGER SHARKS

Australian Ben Cropp has arguably spent more time up close and personal with tiger sharks than just about anyone else on the planet. His shark films are shown worldwide. According to his observations, there is a simple kind of etiquette when tiger sharks dine. It starts out slowly. An orderly queue forms. At the head of the line is always the largest shark. Each individual takes its turn at the prey, and the biggest shark is allowed back in to have an extra bite every third or fourth round. Once into the swing of things, however, all bets are off, and the tiger sharks enter a classic feeding frenzy, occasionally even eating each other.

VASTER AND FASTER THAN EMPIRES

You could be forgiven for baulking at the notion of a shark the size of a whale — up to 14 metres long. That's so much shark it's hard to visualise, but just to give you a clue, the spiracles of the whale shark (*Rhincodon typus*) are so big that small companionable fish swim in and out of them as if they were some sort of underwater cave system! (The spiracle is the first gill slit, which is usually the smallest one. In most bony fish it is closed.) The aptly named whale shark has, in fact, become quite a draw card and several ecotourism ventures offer eager visitors the opportunity to dive and swim with these magnificent, vegetarian animals. (Whale sharks are not to be mistaken for whaler sharks, however. Whalers are carnivorous and definitely dangerous to humans.)

Not all sharks are huge. At only 19 centimetres long, the dwarf lantern shark (*Etmopterus perryi*), for instance, is about the smallest shark you can find, and there are a number of other 'bonsai' sharks. You'd have to be a dedicated shark phobic to have nightmares about something called the pygmy ribbontail catshark (*Eridacnis radcliffei*), which at only 20–25 centimetres long is another of the smallest of all known sharks. The name sounds like something you could market as a toy, a sort of underwater My Little Seahorse. The Caribbean lantern shark (*Etmopterus hillianus*) is a similar size. Just because a shark is small, though, doesn't mean it's nothing to worry about. The cookiecutter shark (*Isistius brasiliensis*) may only be 45–50 centimetres long but it comes from the scarier depths of town, more likely to be occupied by gremlins than My Little Seahorses. This bioluminescent creature lives in the deep but surfaces at night to graze on shallow-dwelling species. It used to go by the relatively innocuous common name of cigar shark, but its razor sharp teeth can and do cause serious damage to far bigger animals, including dolphins and humans. Damage to people is mostly post mortem, but there is some evidence of attacks on living swimmers, which could end up being death by a thousand bites.

BUBBLES AND BARRIERS

Whoever penned that old vaudevillian number 'I'm Forever Blowing Bubbles' might have been in the business of constantly repelling sharks. Apparently certain shark species are not overly fond of bubbles, which some scuba divers have used to their advantage in keeping a safe social distance from their oceanic hosts. *Jaws* author Peter Benchley took advantage of this fact after he got into a bit of a tight spot with his wife and son, because several sharks were expecting a handout of food that the family did not have. Putting out bubbles from their aqualungs, the Benchleys resurfaced in a tight group with the boy in the middle. Fortunately, it worked for them. Unfortunately, no one has been able to turn this particular piece of information into serious protection for beachgoers, although some have tried to create experimental bubble barriers to keep sharks away.

The most reliable shark barriers are the traditional shark nets. In some instances this refers to the complete enclosure of a relatively small space, but more frequently it is a series of nets attached to anchors and buoys. This kind of netting attracts sharks, which are then entangled and drown, as are other large marine species. Exclusion nets, the preferred option for Hong Kong's swimming beaches, are better because they do not result in the deaths of sharks, turtles or dolphins.

A DENTIST'S DELIGHT

Shark skin is rough enough to do serious damage to humans, so rough that in the past it has been used as a sort of organic sandpaper. Shark skin is made up of dermal denticles — small 'skin teeth', so to speak — that serve to aid smooth passage through the water as well as protecting the shark from abrasion and parasites. It is, however, the fearsome jaws full of razor sharp teeth that really focus human fear. The average human thinks about sharks' teeth in one of two ways: terror and lust. By far the most common is terror of the sheer number and sharpness of them. This is understandable. Seen up close, some of those front teeth in the mouth of a great white are not just large and sharp — they are *serrated*. Great whites are, after all, the largest and most enduring carnivorous fish (a success story only recently compromised by humans). They have had many eons to evolve their survival strategy.

Very few people get beyond the basic responses to sharks' teeth, but those who do have some enchanting facts to report. For a start, sharks have a long, long history, having first appeared in the Silurian and Devonian periods, 400 million years ago. Not surprisingly, therefore, sharks' teeth are one of the most plentiful and common types of fossils. Many prehistoric sharks are known to us from a single fossilised tooth. And some of them belong to unimaginably large animals. Place the fossilised tooth of the aptly named megalodont (huge tooth) next to one from a modern great white and it makes the latter look like a miniature baby tooth.

Prehistoric or modern, shark teeth are discarded and replaced regularly. No needle, drill nor amalgam needed. All those serried and serrated ranks behind the front row are just waiting for their chance to move forward and replace any tooth that is lost. And sharks' teeth move in the gum — they can bend and flex easily. Anyone who has lost a tooth biting down on a boiled lolly has reason to be envious.

Most remarkable of all — given that mouth of dental armaments — great whites frequently use their teeth to test out the world, in much the same way as human babies do, or dogs that soft-bite or gum their owners playfully. Australia's Valerie Taylor (see the Damascus Awards on page 116) can testify to having experienced this first-hand. But then she was encased at the time in the flexible chain mail diving suit designed by her husband Ron. I can't imagine anyone else possessing the nerves of steel necessary to hold still for such an examination, even in a chain mail suit, but it makes more sense of the fact that a great number of people who are 'attacked' by great whites actually survive. In reality, if a great white seriously intended to kill you your chances of survival, barring miracles, would be zero. Sharks' teeth can exert a pressure of more than 25 kilograms per square millimetre at their tips. Those who survive a shark's bite have most likely been subject to a check-it-out kind of nibble and had the great good fortune of being spat out as tasting no good or being of no immediate interest.

From a shark's perspective, people are unpalatable really. And even the most obese among us are hardly fat enough to be worth the bother. A large carnivorous shark (they don't all fit that description) would far rather dine on its traditional die: a plump seal or sea lion. It is deeply unfortunate, but hardly the sharks' fault, that seen from below surfers on surfboards resemble nothing so much as a seal or sea lion. (Not all scientists agree with this hypothesis, but having seen an image of a surfer at sea taken from below, it seems plausible enough to me!) This is likely, then, to be one reason surfers feature a lot on the list of shark bite victims and survivors. The other inarguable reason, of course, is that by virtue of their hobby or profession, surfers spend a great deal more time in the ocean than the average citizen.

FAMILY HERO

It is a truism that extraordinary circumstances can sometimes call forth extraordinary acts of strength or heroism, especially when people are protecting their young. One hears tales of petite mothers lifting vehicles to free their trapped toddler. But even when the relationship is not so primal, sometimes protective instincts provide singular acts of courage. When eight-year-old Jesse Arbogast had the misfortune to become the poster boy for the so-called year of the shark attacks (2001), his uncle wrestled the bull shark (more than 2 metres long and weighing more than 100 kilograms) to shore in order to retrieve his nephew's severed arm. With the assistance of a park ranger and firefighter who were at the scene, the boy's arm was saved and, remarkably, surgeons were able to re-attach the limb.

UNSYMPATHETIC MAGIC

It's a very old story that what we fear we wish to possess, hence the enduring fetish for shark tooth jewellery. A reasonably large subset of humans is keen to possess sharks' teeth. A shark's tooth is proof of their macho capacities, or perhaps a fetishistic token to avoid ending up being eaten by a shark. A set of jaws is even better, pretty much likely to appeal to the kind of person whose atavistic taste in decorating runs to stuffed and mounted animal heads. This, in turn, connects with another 'shark' species — the market shark. Where there is demand — for jaws or teeth — there are always those willing to supply them. Not to mention shark cartilage, shark liver oil, sharkskin boots and whatever else can be bought and sold. Perhaps the most tragic of all such transactions is the wholesale slaughter of sharks whose bodies are left to rot once their fins have been hacked off to cater for one of the cruellest trades. Worse still, the fins are often removed from sharks that are caught in trawlers' nets, and the animal is tossed back into the ocean alive. Aspirational people in their millions have been convinced that shark fin soup is some kind of gourmet status symbol. This is not a matter of subsistence or survival eating. It is a consumerist-driven conservation disaster.

Sharks may not be beloved by humans, but like all animals, including people, they have the right to live free of persecution and torture. Even those who are deaf to claims about intrinsic rights might recognise just how useful sharks are in the ecosystems of the ocean. As apex predators they serve an essential function in maintaining the balance of the complex ecosystem they inhabit. If even that doesn't amount to a hill of beans in some people's estimation, then it should at least be acknowledged how much more important sharks are to the human race alive rather than extinct. Some argue that the new ecotourist phenomenon of shark diving substantially raises the market value of a live shark over a dead one, although I'm a little wary of basing conservation on such premises. Tourism trends are too easily divertible on to 'the next great adventure'. But there are other more enduring ways that conserving sharks could contribute directly to human welfare. They have been around for more than 400 million years and have evolved a pretty substantial repertoire in the survival game. Their immune systems are virtually bullet proof, evidently unassailable by the plagues that beset us — cancer, diseases of the circulatory system and infections. Their teeth are impeccable. Imagine what ethical medical and dental research might be able to uncover given the chance. Imagine what it would be like to think differently about sharks.

THE HAWAIIAN OPTION

Unlike snakes, spiders and scorpions, for the most part there seems to be no duality in the symbolism humans accord sharks. No capacity for craft, no creation mystery, no symbol of healing: the secret life of sharks is frequently rendered by two unattractively coupled words, 'mindless ferocity', with 'cruelty' running a close second. Other than an occasional sporting team, few people wish to ally themselves with these alleged qualities of sharks through the sympathetic magic of adopting the name of a creature so universally despised and feared. There is some debate these days about whether animal species other than humans can possess moral capacities, such as altruism or malfeasance. Whatever position you take on that matter, predatory species in the natural act of preying can hardly be construed as cruel.

Sharks remain as enmeshed in human myths as they are in human drift nets, and few such myths are redemptory in any sense. But there is one. Hawaiians, those original surfers, are unique in holding sharks in high regard, while giving due respect to the danger they present. For a Hawaiian, any shark they meet might possibly be related. Known as an *aumakua*, the reincarnated former relative would look out for the family's interests. Do not be too ready to dismiss this as primitive superstition. At a deeper level perhaps the Hawaiians are the only group of humans to have recognised the fundamental truth that, like it or not, we are kin to all the animals which share our planet, most of which is ocean.

THE DAMASCUS AWARDS

In the shark category, it was impossible to pick a single winner for the Damascus Award. Nor should it be considered parochialism that all three winners are Australians. Take a quick look at the credits on any internationally released shark footage for the last 45 years, including documentary and feature films and television, and the same three names are prominently featured. Jacques Cousteau and Hans Hass may be the pioneers of the underwater film, but Ron and Valerie Taylor and Ben Cropp can collectively be credited with existence of the subgenre of 'shark films'.

The stories of these individuals are very revealing, not only of their private enthusiasms and changes of heart, but also the distinct shift in public perceptions of sharks that all three have contributed to. Key aspects of their stories share a common theme. All three have witnessed first-hand the enormous beauty and diversity of underwater life and the extraordinarily rapid demise of that beauty and diversity over their own lifetimes. To begin with all three were keen spear-fishers — or spearos — winning many championships and trophies in that sport. During its heyday in the 1950s and '60s, the general premise was to kill and catch as many fish as you could within the competition time. And they excelled at it.

Although he and the Taylors have had different professional trajectories since then, Cropp worked together with Ron Taylor in the very early days. At the time of their first film, *Shark Hunters*, Taylor was the cameraman and Cropp the hunter. It was the first time the champion spearo would kill a shark. Over the next four years he would kill hundreds. A meeting with the sublime whale shark in 1965 caused him to change his mind about the legitimacy of killing sharks for spectacle. And, despite a close encounter with temptation in the post-*Jaws* frenzy, he has stuck with that decision.

Valerie Taylor has won Australian women's spear-fishing championships. As well as many Australian men's championships, Ron Taylor also won the world championships in Tahiti in 1965. It must have been a watershed year, because it marked the turning point for him. He says: 'Giving up spear-fishing was the best thing I ever did.' Now he prefers to hunt with his camera. Valerie agrees. 'After about ten or fifteen years of spear-fishing competitions and spear-fishing just for the table, we realised that they [the fish] weren't coming back. We had to go further out, we had to go deeper. And we were actually wiping out all the big, edible fish.'

Rod's hunting with his camera has left the world a staggering legacy of shark footage, including the very first footage of the great white shark filmed outside a cage, and sharks filmed by night. Most unforgettable of all are images of Valerie being 'mouthed' and 'chewed' by curious sharks, while protected with the chain mail suit Ron designed and had manufactured. The element of showmanship, the echo of carnival, is ever-present in their work: the eye for spectacle is undeniable. Yet Ron and Valerie Taylor and Ben Cropp have shared with the world what they have seen. And they have not been afraid to change their minds and habits and put their considerable talents and expertise into protecting the animals they once hunted. This is doubly admirable given that when they took these pioneering stands, it was at a time when conservation was barely heard of in public discourse, and the old dictum 'The only good shark is a dead shark' was unquestioned.

All three are first-rate underwater photographers. All three are famous for promoting public understanding of a much feared animal.

> All three have won the Order of Australia for their services to conservation, as well as other prizes and awards. Now, all three have earned the humble Damascus Award.

5

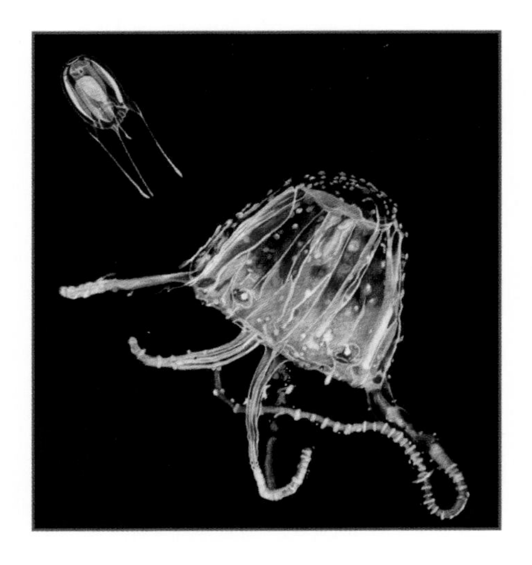

Beach and Beyond

Cone shells, blue-ringed octopuses, stonefish and box jellyfish

If you are predisposed to phobias about the deadly and dangerous animals of the deep, no doubt it's the marine megafauna that feature in your nightmares. But wait, there's so much more to be anxious about. Taken collectively, the smaller sea critters of reef and shore have a far greater impact on humans than sharks or rays. As usual, humans as a species pose more of a threat to all of these species than vice versa. Also as usual, the Anti-Tourist Board has been at work enlarging and elaborating the extent of the threat posed by box jellyfish, blue-ringed octopuses, stonefish and cone shells. Still, it's quite true that these kinds of animals can and do cause an enormous amount of human pain and suffering, even if they do not take a high toll in terms of human deaths. Fortunately, sensationalist press notwithstanding, encounters with these dangerous animals are largely avoidable. None of these species seeks us out so let's go in search of them within the absolute safety of this book.

HANDSOME IS AS HANDSOME DOES

Tulip cone, cloth-of-gold cone, cat cone, leopard cone, monk cone, glory of the seas cone: some cone shells have singularly enchanting names. Others are less attractive: flea cone, rat cone and spiteful cone. The names seem to offer a before and after parable of human relations with this animal. Many species of this genus are highly prized by shell collectors for their beauty and in some cases rarity. 'Spiteful' is perhaps the least of the negative epithets that might be uttered by any unwary person stung by one of the eighteen species of cone shells that are definitely dangerous to humans. According to the indefatigable Australian toxicologist Struan Sutherland, seven species top the list: geographer cone (*Conus geographus*), glory of the seas (*Conus gloriamaris*), feathered cone (*Conus omaria*), striated cone (*Conus striatus*), textile cone (*Conus textile*), princely cone (*Conus aulicus*) and tulip cone (*Conus tulipa*). And, of these, it is the geographer cone that is implicated in the most human deaths. The other 280 or so species of cone shell have some venomous capacities so they all deserve cautious respect, but none are so distinguishably dangerous as the magnificent seven.

Yet even these seven are not major players in human fatalities. Worldwide the reported death toll is variable, but totals always number in the tens or dozens rather than the hundreds or thousands. On the other hand, there are much higher numbers of people who survive such stings, and many of them experience quite a lot of morbidity (ill health) and suffering. At the time of envenomation by a cone shell a person will usually feel sharp pain, which can range from mild to unbearable. Some stings leave no mark, others a whitened area with a bluish hue. Numbness and swelling are common. With serious stings, there is a range of distressing symptoms including sensory disturbance, severe itching, paralysis and brain swelling. Coma and cessation of breathing presage death unless skilled and sustained CPR is continued until the person can be attended by emergency medical personnel.

THE BIGGEST SHELL

Cone shells are marine gastropods: snails in lay language. And — as even preschoolers can observe — snails carry their home on their back. But the insatiable desire to collect the biggest shell, the prettiest shell, has led to a worldwide trade that threatens the survival of these gorgeous creatures. The collecting and buying of shells is nothing new; the European shell trade has been in existence for at least the last 500 years. A cone shell auctioned in Amsterdam in 1796 fetched a higher price than a Vermeer painting.

What seriously concerns many eminent scientists today, however, is the sheer scale of the trade (millions of animals are killed annually), the lack of regulation in most countries and the concurrent threats to cone shells from environmental destruction. After all, as for all of us, a snail's true home is its habitat. For cone shells this is frequently mangroves (half of which have already been lost) and coral reefs (a quarter of which are seriously damaged). Both habitat types are suffering serious and sustained threats ranging from pollution, through predator species such as crown-of-thorns starfish, to global warming.

Grave fears for the loss of species, along with their potential for human benefit prompted Eric Chivian from the Harvard Medical School and two colleagues to publish a landmark letter in *Science* in October 2003, urging scientists to take these threats seriously and take action to help conserve cone shells.

SNAIL MEDICINE

Cone shell toxin has more than 100 different constituents, each with variable effects from paralysis to pain. Like those of many other marine invertebrates, cone shell toxins are biologically active agents that can potentially benefit humans. Some of the possibilities are antibiotic, antiviral, antifungal and anti-inflammatory compounds. This represents a veritable pharmacopoeia, at least potentially.

This potential is being followed up by active research into the biological pathways of cone shell toxins that may end up providing far more effective analgesics for nerve pain in cases where opiates are inadequate or ineffective. This could provide great benefits for people who have some complications of diabetes, certain forms of cancer and AIDS — all of which can produce excruciating and virtually untreatable pain. Already there are stage-three clinical trials for a synthesised drug called Prialt, developed from cone shell research. Prialt is estimated to be up to 1000 times more effective than morphine, and without its addictive potential.

Research into cone shell toxins requires a supply of cone shells. Many of these are harvested from the wild, leading to some concern that this additional stress on an animal species that is already over-collected for the shell trade may well result in extinctions. One way around this is to farm live animals, which is the approach taken by research professor Jon-Paul Bingham. Bingham is an Australian who works from the Laboratory of Molecular Neuroscience at Clarkson University, Potsdam, New York. He is the author of many scientific papers, including the intriguingly titled 'Cone shells, condoms and cod liver oil — The study of the milked venom from toxic carnivorous snails'. The introductory phrase sounds like something from the sleazier side of the sex trade, but of course it's not. Apparently, unlubricated condoms are the perfect membrane for milking cone shells of their venom. Now that's practising safe science!

BEYOND BLUE

It may sound like an urban myth, but it lacks the macabre ending and is in fact true. A friend of mine went for a walk down to her local swimming beach in Western Australia one summer and came across a group of tourists who — judging by the depth of their sunburn — were on holidays from Britain. They were gathered around a bucket peering in animated fascination, and gently poking the occupant of the bucket with a child's spade. In the few inches of sloshing seawater was a small, sandy-coloured, seemingly innocuous little octopus. In fact, a very upset and angry little octopus. Such a tiny creature; it looked so lovely when mad, its thin, peacock-blue rings pulsing.

'For pity sake don't touch it!' 'They're deadly!' 'Their sting is agonising!' The Aussie onlookers — including my friend, who knew that the last claim was untrue — were well aware how implausible their warnings would sound. To non-local ears it could so easily and perilously seem like a classic tease. Anything less resembling deadly giant octopuses, squids and ancient mariner's monsters would be hard to imagine. Fortunately for them, the group collectively conceded that discretion was the better part of valour, and declined to empirically 'disprove' such seemingly fallacious advice.

But what if they hadn't? What if they had gone ahead and picked up the animal? Well, the blue-ringed octopus is definitely small: body size is usually from 4–6 centimetres and tentacles vary from 7–10 centimetres, thus the span of its tentacles is unlikely to be much bigger than 20 centimetres. However, it is absolutely true that they have venom capable of killing a human. In this case, size really does not matter because the blue-ringed octopus is considered the most venomous genus of octopus in the world. There are around ten species in the genus. Two of those found in Australian waters have been known to cause fatalities. At first, they appear to be named with the typical Australian linguistic reversal: the larger animal is called the lesser (sometimes southern) blue-ringed octopus (*Hapalochlaena maculosa*). Its slightly smaller cousin, *H. lunulata,* is called the greater blue-ringed octopus. However, the 'greater' and 'lesser' refer to the size of the rings, not of the animals themselves.

Members of the blue-ringed octopus genus can live in water from 50 metres to extremely shallow shorelines, anywhere from the Sea of Japan south to Australia and across to Vanuatu, the Philippines and Sri Lanka. For example, *H. maculosa* is only found in southern Australia but occupies the whole vertical span of the continental waters. *Hapalochlaena lunulata*, on the other hand, has a broader distribution but does not live at such depths, being only found down to about 20 metres. *Lunulata* are often found in rock pools and tidal flats, areas also frequented by humans. It is not uncommon for them to take refuge in small containers both natural (shells empty of their original inhabitants) and artificial (tins, bottles, etc.). Given the widespread distribution of these species in the subtropical and temperate waters, and their potential toxicity, it's surprising that relatively few people have actually died (two in Australia and one in Singapore over the last 50 years, although many more have been bitten, with variable consequences). Perhaps this is largely attributable to the fact that the species is singularly non-aggressive, and not likely to attack unless severely provoked — by being confined in a bucket and prodded, for instance. Also, despite their wide distribution they are considered to be relatively rare across that range.

Another factor reducing the risk is that blue-ringed octopus venom does not come from tentacle stingers nor is it injected. It comes from glands near the animal's small beaked mouth, and flows in with its saliva when it bites, generally for predation (crabs are a favoured food) — and occasionally for protection. It's not unheard of for people to be bitten under the water; those who are have some advantage as it may mean less venom actually enters the wound. Overall, however, a person would have to be extraordinarily unlucky to get randomly bitten: the majority of recorded bites occur when people pick up what seems like a pretty and harmless sea creature and put it on their outstretched arm or hand for a closer look.

So if one of the tourists had picked up that little blue creature and been bitten, what would the outcome have been? The blue-ringed octopus's 'poison of choice' is tetrodotoxin, a neurotoxin that induces swift paralysis. This is the same toxin that is found in the flesh of toadfish and pufferfish. Blue-ringed octopus venom also contains hapalotoxin, although this is not considered much of a problem either for humans or crabs. Its chief role seems to be as a means of defence against fish that prey on the octopus. Around 7 milligrams of venom is a common yield for the average sized blue-ringed octopus, which is sufficient to paralyse ten adult humans. Unlike the stonefish or stingrays (whose venomous arsenals feature a different array of chemicals), blue-ringed octopus venom does not cause much pain if any. Even the physical pain of the bite may go unnoticed, especially with underwater encounters. In fairly short order, the bitten person would experience tingling and numbness of the face and neck. Muscle weakness, sight and speech disturbances follow. There is often difficulty breathing and occasionally vomiting. Finally, paralysis ensues, and this may last from four to twelve hours. Particularly vulnerable people — children, elderly or frail individuals — can die within half an hour, so swift first aid and urgent medical attention are essential.

MADONNA CEPHALOPOD

The blue-ringed octopus is the ultimate in mother martyrs. Once she has been inseminated via a male's modified tentacle (the hectocotylus) she awaits the right time and conditions to release her eggs. She then cares for them by guarding against predators and occasionally oxygenating the eggs with jets of water. During this time she does not eat. Once the eggs are safely hatched, she swiftly declines and dies. The lifecycle of the blue-ringed octopus is two years.

TOUCH NOT THE OCTOPUS

My ancestors' clan motto in Gaelic is *Na bean don chat gun làmhainn*. It means 'touch not the cat without a glove'. The same applies to blue-ringed octopuses, although people don't have much call to worry about them in Scotland! (Not yet, anyway — see 'Indicator species?', on page 136). However, the motto is apt and should be adapted to say 'touch not the octopus at all!' Sadly, however, the old story of 'what can be bought and sold will be bought and sold' applies even to this little creature. The trade in blue-ringed octopuses for aquariums is a stupid idea. Ask Dr Roy Caldwell, a researcher from California with a professional interest in cephalopods. Even with all his professional knowledge and skill he has very nearly been bitten by a captive blue-ringed octopus. He presents some compelling reasons for the home aquarium owner to leave blue-ringed octopuses where they belong — in their own environment. Primarily, this is because the species cannot sustain such large-scale capture simultaneous with major threats to its habitat. Even from the consumer's point of view it doesn't make much sense to buy (and re-buy and re-buy) an expensive little animal that keeps dying on you, unless it gets to kill you first. They are, after all, deadly. Why risk it?

A WORSE FATE

Blue-ringed octopuses, cone shells and stingrays are all possessed of a singular beauty. The best you can say of a stonefish, alas, is that it is unlikely to cause death if it stings you. The kicker is that you might want to die, as the pain of a stonefish sting is excessively excruciating. The effects are well known to traditional coastal dwellers and they are dramatically enacted in the corroborees of the Aboriginal people of some northern Australian groups. Practitioners of traditional medicine from northern Australia to coastal Africa have developed remedies to address the pain caused by marine stings from stonefish and stingrays, some of which are in current use. Their knowledge is also reflected in the precise clinical description of the effects of stonefish envenomation that shines through the dialect of this Australian Aboriginal man speaking more than 100 years ago:

> *Suppose that fella nail go along your foot, you sing out all a same bullocky all night. Leg belonga you swell up and jump about. Bingie [belly] belonga you, sore fella. Might you die. (quoted in E.J. Banfield's* Confessions of a Beachcomber *1908).*

Well might you wish to die, but current medical thinking is that it is extremely unlikely for a person to die from a stonefish sting. The cause of the few deaths that have occurred in the immediate days following stonefish envenomation is perhaps attributable to secondary infections, including tetanus.

DOCTRINE OF SIGNATURES

Those of a mediaeval mindset might take note that stonefish have *thirteen* venomous spines: an ominously obvious number for something so scary. The thing about stonefish, though, is that they are the ultimate practitioners of passive aggression. Researchers quite specifically believe that stonefish and their relatives evolved such a fearsome array of venoms together with a completely unobtrusive appearance in order to be lazy. They are marine couch potatoes, moving only when absolutely necessary, and firmly believe in eating home-delivered.

The stonefish came by its common name because it looks like, well, a stone: knobbly, often greeny-grey and covered in warty bits and algae. They are highly unlikely to win a beauty contest, that is if you ever get to see one. Stonefish are plentiful but highly camouflaged, and even those who are stung seldom see the administrator. Of the ten species in the stonefish family (Synanceiidae), two are most referred to in the medical literature: *Synanceia horrida* (did the namers mean its looks or its effects?) and *Synanceia verrucosa* — the reef stonefish. *Verrucosa* means warty. All these are somewhat mild monikers compared with the names bestowed on some of the scorpionfish, which are fairly close relatives of the stonefish, belonging together with them in the Order Scorpaenoidea. One genus is dauntingly titled *Inimicus*, and the common names include bearded ghoul, devil fish and demon stinger.

A complete ring-in is the bastard stonefish, which is not even remotely related. It is a common name sometimes bestowed on the non-venomous frogfish that belongs to a different Order of fish altogether.

INSIDE AND OUTSIDE THE BOX

Stingrays may have involuntarily hogged the headlines in recent years, but their dangerous aura is somewhat of a chimera. It pales into insignificance when compared with the danger posed by the see-through jellyfish that share their watery kingdom. All 9000 Cnidarian species are venomous. No other phylum (large group) in the animal kingdom matches them. The word Cnidia means 'nettle', which seems like a bit of an understatement when it comes to the group of animals that go by this name. Every one of them is a champion stinger, but one genus stands out as the premier problem for humans. The hands-down winners when it comes to venomous marine animals are the box jellyfish, sometimes called sea wasps. There are around twenty species of box jellyfish. And of these the single most dangerous species is *Chironex fleckeri*, which has earned itself the rare poetic common name of 'fire medusa'. Some scientists believe that this relatively passive animal is, in fact, the most venomous species in the world, and without doubt the most venomous marine animal.

Death can occur within minutes of receiving a sting. The death-dealing mechanism is a structure called a nematocyst, a word often used synonymously with Cnida, the stinging capsules that distinguish these animals. (There are, in fact two other types of Cnida — spirocysts and ptychocysts — but nematocysts are the only ones capable of puncturing the skin.) The structures can also be found in some corals, sea anemones, hydroids and gorgonians, as well as jellyfish. One of the reasons *C. fleckeri* excel as venomous animals is because they possess several different types of nematocysts designed to puncture, hook or exude a substance that sticks to human skin. The various operations of *C. fleckeri*'s nematocysts could genuinely be described as 'overkill'. The scientific consensus is that the swiftness of death in fatal stinger episodes must be dependent on direct injection of venom to the blood vessels and hence into the bloodstream. Children are the most likely to die from jellyfish envenomation, although other factors that increase risk are a person's thickness of skin, the size of the jellyfish and the number of nematocysts released.

For humans as a species it would be easy to take it personally, but of course these mechanisms for prey capture evolved quite independently of us. We are of no use to jellyfish as food, as we are not their prey. This is quite different to larger deadly carnivores such as crocodiles, lions, sharks and bears which can and do eat those they kill, given the chance. Anyone unlucky enough to be stung by a jellyfish serves no such function, but is simply collateral damage of the animals' extremely efficient mechanisms for capturing their actual prey, which are small fish and crustaceans. *Chironex fleckeri* may be a passive hunter like the rest of its transparent and spineless relatives, most of whom rely on tides for transportation and bringing prey in range. However, because the species possesses extremely sophisticated eye structures, some scientists hypothesise that hunting may be an active process. Either way humans are not the target.

In other words, then, they won't come looking for you. The best idea is to simply stay out of their way. The current distribution of stingers is generally accepted to be north of the Tropic of Capricorn, throughout the Indo-Pacific region (see 'Indicator species?' on page 136 for news of potential change). Those who live or travel to these areas should take heed of stinger season, which pretty much follows the wet season, anywhere from October to May. Of course, jellyfish pay attention to environmental cues rather than calendar dates so these are approximate parameters. The closer to the equator the earlier they appear, as early as August along Australia's north coast. Popular public beaches on the east coast of northern Australia are patrolled and protected in a number of ways, including netting. But that is hardly practical for the vast areas of coastline of Indo-Pacific countries. In this case, local knowledge is invaluable. As a general principle, however, certain spots are known to attract congregations of jellyfish, including creek and river mouths — any estuarine habitat is favoured for generally hanging about. These areas are also used for spawning in late summer. Stingers are less partial to surf, coral reefs and deep water, but this is not a guarantee of their absence in such conditions. Swimmers beware! If you simply must swim, then a stinger suit is *de rigueur*. Locals would undoubtedly do well to invest in one, but improvising is always an option to improve your odds (see 'The peculiar powers of pantyhose' on page 137).

THE INTERNATIONAL CONSORTIUM FOR JELLYFISH STINGS

The history of jellyfish stings has tangled tentacles of woe, confusion and pain. People were stung and died, and it was often unclear what had killed them. There was also a lot of taxonomic confusion about various jellyfish species among scientists. Knowledge of first aid was scanty, often misleading and myth-ridden. A few individuals worked hard and pioneered solid, reliable data about certain species but the general picture was not very clear. This all started to change in the 1980s when two Australians, Dr John Williamson and Dr Peter Fenner, began to work systematically, drawing together the existing expertise, sorting out the unreliable stories and building a robust central data bank. They co-founded the International Consortium for Jellyfish Stings in 1989. One of the first things that became evident was that although Australia is one of the world's hotspots for jellyfish envenomation it is by no means the only place to have such problems. The consortium followed up data from around the world drawing their information from a wide variety of sources, including media, medical and surf lifesaving reports, as well as toxicological and marine scientists.

Best of all, from modest beginnings of research and publications originated by the Queensland Surf Lifesaving Club, Williamson and Fenner — together with Joseph Burnett and Jacquie Rifkin — have developed a comprehensive and peerless international medical and biological handbook entitled *Venomous and Poisonous Marine Animals*.

INDICATOR SPECIES?

Environmentally literate people are familiar with the concept of 'indicator species': animals or plants so highly sensitive to changing conditions that their response to such change is swift and obvious to observers. Indicator species function as early warning systems of potential environmental damage or collapse. The classic example is an amphibian, such as frogs. In these days of global warming, many more species are being added to the list of indicators and jellyfish are one.

In late 2007 a plague of 'mauve stinger' jellyfish wiped out Northern Ireland's organic salmon fisheries. Writers of scripts for C-grade schlock-horror movies take note: the 'mauve stingers' en masse covered an area of 26 square kilometres to a depth of 11 metres. This has not been a long-standing occupational hazard for Northern Ireland's fish farmers. The 'mauve stingers' were thousands of kilometres north of their normal preferred habitat. This is a startling, but not singular event. There are plenty of other data to suggest that as the climate shifts so do the stingers.

PERSPECTIVE ON POISONS

Dr Peter Fenner, co-founder of the International Consortium for Jellyfish Stings, notes all known deaths from these marine animals, which gives a relative indication of their danger:

Chironex box jellyfish (67)
Irukandji jellyfish (2)
blue-ringed octopus (2)
stingray (2)
cone shell (1).

THE PECULIAR POWERS OF PANTYHOSE

Some men but almost no women mourned the passing of old-fashioned stockings. Some people, however, saw the potential value of the new artefact — pantyhose. Plenty of bush mechanics have used them to substitute for a broken fan belt. But Dr Jack Barnes, a Queensland radiologist, really knew how to think outside the box. He was responsible during the 1960s for Queensland surf lifesavers wandering around unselfconsciously with pantyhose on their legs, and also (with a strategic snip at the crotch) pulled over their heads and arms. Even such a thin barrier was enough to make it safe for the lifesavers to enter the water and net, looking for stingers at public beaches. When they found them, they closed the beach. These days, purpose-designed and more attractive stinger suits are commercially available, but the principle of protection remains the same.

Barnes, along with Ronald Southcott and Hugo Flecker, is one of the pioneering heroes of stinger research. He is memorialised in the scientific name of the jellyfish that cause Irukandji syndrome: *Carukia barnesi*. He earned it. Nobody knew what caused the so-called Irukandji syndrome before he lay for hours in shallow water, breathing by scuba, until an insubstantial and transparent animal swam across his field of vision. Having trapped a couple of the suspect jellyfish he proceeded to sting himself, his son (presumably an adult) and a surf lifesaving friend. Sure enough, all three manifested the symptoms of Irukandji after 30 minutes, and ended up in hospital in severe pain. So his name lives on. History does not record the thoughts of the anonymous friend, the son or the son's mother. By all reports, Hugo Flecker was an equally excellent field naturalist, although not given to such extreme methods. His name is honoured in the scientific name of the scariest jellyfish of all — *Chironex fleckeri*.

There is no Damascus award for this chapter, although it could possibly be argued that brave Jack Barnes was a potential candidate for a Darwin Award! Nonetheless, the knowledge gained by him and his colleagues has made it far easier for humans to co-exist much more safely with these deadly marine animals.

6

Pack Drill

Wolves, dingoes and coyotes

The hound slumbering by the hearth may seem the very essence of domestic bliss in a safe home secured from the tricksy coyote. This tame beast is the trusted domestic guardian. And although the wolf no longer turns up at most doors, the dog's wild instincts sound the alert to other intruders. Such wild instincts are not so easily switched off, even after centuries of domestication. That's why 'man's' best friend is frequently 'man's' worst enemy, at least among members of the Canidae clan. Each year dozens of children, and a smaller number of adults, die from domestic dog bites, according to international statistics (domestic cat bites are also a problem, but that's another story). As always, numbers of cases go unreported, particularly in developing nations, which have a surfeit of death from all causes and a deficit of infrastructure for accurate record-keeping. However, since the addition of 'E code 906.0' to the International Classification of Diseases, dog bite fatality numbers in developed nations are clearly evident. Acquiring a specific code beyond the generic 'Other' is a sign in itself that death by dog bite registers as statistically significant. Still, the mortality figures are not particularly high. Reality really sinks its teeth in when you see the morbidity figures. For example, in the United States for each dog bite death there are estimates of close to 700 hospital admissions and 16,000 visits to emergency centres. This represents a huge amount of trauma, both mental and physical, and in extreme cases, permanent disfigurement.

Still, for many people it's a case of 'better the dog you know'. For that reason alone, it's worth becoming a bit more familiar with the demonised wild species we are so wired to fear. Most people could easily list several of the domestic dog's wilder cousins — wolves, dingoes, coyotes, foxes and jackals. Apart from aficionados of David Attenborough documentaries or the Discovery channel, fewer people are familiar with the less common species of wild dogs, such as the dhole, or Asiatic wild dog, and the remarkable African wild dog that has ears like twin satellite dishes and hearing to match. (Neither of the latter animals presents much of a threat to humans, although the reverse is far from true. Both are severely compromised due to direct and indirect human-caused factors).

All the animals listed above are classified in the family Canidae. In common parlance, these are frequently divided into two sub-groups: dogs and foxes. The taxonomic reality is far more complicated, but it's a reasonable enough division for current purposes. Overall, the shadow cast by the wild dogs is far larger than the actual threat they present. Nonetheless, realistic information about the dangers brings perspective which in turn gives birth to respect.

THEY ALSO HOWL WHO ONLY EAT CHICKENS: FOXES AND JACKALS

There are 27 species of fox in the world, and none of them poses a real danger to humans. This much-maligned animal, not much larger than the domestic cat, has been persecuted, hunted and trapped for sport and trade. They will, for certain, kill chickens and possibly other small domestic animals that are unsecured. They might even bite a person if cornered, but are far more likely to simply run away, given the opportunity.

There are three or four species of jackal, depending on the source you consult. Although jackals do not prey on large animals such as humans, they still can threaten human life. According to a report in Sri Lanka's *Sunday Times Online* the villagers of Ali Oluwa are being besieged by rabid jackals, a phenomenon that is not just confined to Sri Lanka. In this instance the culprit is a golden jackal (*Canis aureus*), a species with a very wide distribution throughout much of the world. Many jackal populations are reservoirs of rabies. For example, south-central Africa also suffers from rabid jackals, although in this case the species is the black-backed jackal (*Canis mesomelas*). It's a significant issue because, according to WHO statistics, the human death toll from rabies is around 50,000 annually. In fact you are a hundred times more likely to die of rabies than you are to be killed by a macroscopic animal of any kind (well, apart from other humans, that is).

WOLVES OF THE NEW WORLD

Throughout the northern hemisphere, across Europe, Central Asia, India and the Americas, the wolf once reigned as one of the most widely distributed mammals. It remains the largest and most feared of all the wild canids. No wonder it is so tightly woven into our cultural fabric. Our shared dreams, nightmares, fables and fairytales are still full of wolves. Our world is not.

The general consensus is that there are only two true species of wolf — the grey wolf (*Canis lupus*) and the red wolf (*Canis rufus*). Grey wolves are the larger of the two, weighing up to

62 kilograms. Of variable size, an average length from snout to tip of tail is around 1.6 metres for males, and slightly less for females. Although they differ in several distinct details, grey wolves are fairly similar to a German shepherd dog in size and general appearance. The red wolf, an American species, is smaller than the grey wolf and larger than the coyote. Scientists are still debating whether *Canis rufus* is actually the result of hybridisation of grey wolves and coyotes or an independently evolved beast. Either way, the major threat to the red wolf is extinction through hybridisation. This is the same problem facing the dingo, only for red wolves the problem is not domestic dogs, but interbreeding with coyotes or coyote crossbreeds.

As might be expected, though, given their widespread distribution, there are a number of subspecies of grey wolves: twelve in total. Taxonomic wrangles mean some subspecies may be lost to reclassification, especially in the United States. At the same time, an argument has been put forward that the wolves of eastern North America are an indigenous species that evolved *in situ*, rather than wandering across from Eurasia sometime in the dim mists of prehistory. The name proposed for this allegedly native species is *Canis lycaon*, but it has not yet been universally recognised as a separate species.

Many subspecies of grey wolf are extinct; others are tiny remnant populations hanging on here and there in pockets of suitable habitat. Its howls are no longer heard through much of Western Europe, many areas of the United States and all of Mexico. Owing solely to their widespread distribution, however, the overall conservation status of grey wolves is designated as being of 'least concern'. This is a relative term. Of course, those who belong to the taxonomic 'clumping' clan don't see too much of a problem, but many people are fighting to keep various populations of wolf subspecies viable in the interests of preserving diversity.

REST IN PEACE

Although children are far more likely to be the victims of wolf predation, adults have not always been considered invulnerable. It seems that in some places not even the dead were considered safe from the depredations of wild wolves. At one stage, the Highlanders of Scotland favoured burying their loved ones on islands to ensure that the wolf was kept from the door of their last dwelling.

THE WOLF WARS

In the United States, the likelihood of being attacked by healthy wolves is extremely remote. Not a single case of a predatory attack by a wolf was recorded for the entire twentieth century, although one death of an adult male was recorded in Canada in 2005. The unfortunate Kenton Carnegie was apparently trailed and hunted by a wolf pack after he left a surveyors' camp in the wilds of Saskatchewan: a terrible, but thus far unique, death. Over the previous hundred or so years there were around 80 aggressive encounters on record, but the majority (62) of the animals concerned were rabid. Of the remaining eighteen attacking animals, eleven were ones that had become habituated to humans. Only six attacks were classified as acts of unprovoked aggression.

Despite this, there has been a long tradition of killing wolves in America. Red wolves have fared far worse than grey wolves, and in the 1980s were designated as extinct in the wild. Perhaps it was the heritage from European immigrants who had a long history and mythology concerning wolf predation. Perhaps it was the fearful reaction of pioneers in a territory that offered so much unseen danger, as well as opportunity. No doubt there were multiple reasons, but a full-scale war was waged against the American wolf, one that saw the animal become extinct across much of its former range.

As attitudes to wildlife underwent a sea-change in the latter half of the twentieth century, wolves were among the flagship species in a new kind of 'pioneering' — reintroduction programs. The red wolf was re-established in the wild in 1987 in North Carolina. The species has regained a precarious foothold there, but at least a quarter of them die on the roads, and the rest are in grave danger of being bred out and so once more becoming extinct in the wild. A decade or so after this first effort, one of the most famous reintroductions of grey wolves to their former range was carried out in Yellowstone National Park. This program has been hailed as an international success story in the management of endangered species, although it has not been without very vocal critics, particularly from those who run livestock nearby.

The wolf wars continue to be fought in North America between groups who wish to continue the protection and those whose most primitive fears are roused by the rise in wolf numbers. On March 2008, the wolf was removed from the federal classification of an endangered species. Exactly four weeks later, eight major environmental groups launched a legal challenge to the wolves' declassification, claiming that it was far from evident that the wolves had established viable and stable numbers that would be proof against a major backlash against the species. It seems that they had solid grounds for their concern. Within that four-week window of opportunity, several states, including Wyoming, enacted legislation to make it legal to kill wolves that were perceived to be threatening livestock. Ten wolves were shot, one of them a member of the Druid Peak pack of Yellowstone National Park.

It's a bitter battle and passions run high on both sides, but there's a fair amount of evidence that the threat of wolf predation on agricultural animals has been wildly exaggerated. Real figures from Montana indicate that the risk of losing sheep or cattle to a wolf is close to zero (0.01 per cent cattle and 0.08 per cent sheep were killed by wolves). And, as mentioned above, no human deaths were recorded in the whole of the twentieth century, so the risk of wolf predation on humans seems miniscule although not impossible. There has been a rise in aggressive encounters in the early years of the 21st century. A few days before Christmas 2007 a family dog headed off some half-starved wolves that appeared intent on closing in on three preschoolers who were tobogganing with their family in northern Canada. Yet even the children's father, Kyle Keays, a licensed hunter who understandably shot the wolves, admitted that it was an anomalous event. For one, according to Keays, wolves aren't known to eat other wolves' carcasses and for another, they aren't known to prey on humans either. 'You never see them. I work out in the bush all the time and normally your first glance is your last glance,' he says (meaning the animal disappears swiftly).

WHEN IS A WOLF NOT A WOLF?

The Ethiopian wolf (*Canis simensis*) belongs to the canid family and, as its common name suggests, it's very wolf-like. However, taxonomically speaking this animal is not classified as a true wolf. Neither is the maned wolf (*Chrysocyon brachyurus*).

LITTLE RED RIDING HOOD

Fear of wolves may be greatly disproportionate in the United States, but Europe, Eurasia and India are a different story. In the last 50 years or so there have been seventeen documented wolf attack deaths in Europe and Russia that are not attributable to diseases such as rabies or self-defence on the part of the wolf. Even this is not an especially high number, but no doubt that is at least partly due to the extinction of the wolves in much of their former European range. For reasons not entirely clear, the grey wolf is definitely more likely to attack humans in areas outside the United States. Some have speculated that the omnipresence of firearms in the United States and the extensive wolf hunting of that nation have contributed to American wolves being far warier of people than their counterparts in other nations. It's a plausible theory, at any rate.

Whatever the reason, historically people who dwelt in woods in Europe had sound reasons for warning their offspring about the dangers of the 'big bad wolf', most especially if said offspring were so foolish as to be wandering about alone with an enticing basket of goodies serving as bait. Forget the Freudian interpretations: wolves eat children. There are reliable records of hundreds of people being killed in predatory attacks by wolves in the eighteenth and nineteenth centuries. One luckless parish in Estonia lost 48 children to wolf attacks between the years 1808–53. Thirty-six Russian children died from wolf attacks in a much shorter period of the twentieth century between 1944 and 1953. High numbers of child fatalities are reported in India also. Estimates number in the hundreds in the Uttar Pradesh and Madhya Pradesh in recent decades. Such deaths usually occur in the remote and poor rural regions, where the people would have only sticks and stones for defence.

THE COMMON ENEMY

There's a perennial, possibly apocryphal, tale of a Christmas ceasefire in the trenches during World War I. A less well-known story from the same war tells how starving wolves so plagued soldiers fighting at Kovno in Lithuania that the Imperial German and Imperial Russian armies had no option but to join forces temporarily to defend themselves against the common enemy.

DINGO

Like many people, I used to think of the dingo (*Canis lupus dingo*) as an exclusively Australian animal. While it certainly is a most iconic member of Australia's fauna, according to Australian scientist, Dr Laurie Corbett, the best place to find a sizeable population of pure-bred dingoes is Thailand! Thailand and the northern half of Australia are two of the last refuges for this wild animal that all too easily hybridises with domestic dogs, and is everywhere vulnerable because of it.

Corbett has devoted a lifetime to studying all aspects of the dingo, a descendant of the wolf tribe, specifically the Indian wolf (*Canis lupus pallipes*). Its natural distribution spread out from India across southern Asia, and from there sailors ensured that it travelled further and took up residence on many southern hemisphere islands, not so much as a favoured companion, but more like a living larder. Dingo and dog are still considered good food in some parts of the world. Sailors were in the habit of carrying other small animals such as rabbits, and releasing them on islands to provide food in case of shipwreck. It is likely that in a somewhat similar fashion the dingoes that didn't end up as dinner eventually arrived on the northern coast of the southern continent about 3500 or 4000 years ago. From there, the dingo again spread southwards, successfully taking up residence in all manner of habitats, only stopping short of crossing Bass Strait to Tasmania. It met with varied fortunes among different indigenous nations, but was frequently valued as both a hunting companion and an animate hearth rug, hence the expression 'three-dog night' for exceptionally chilly weather. Among some indigenous groups the dingo even achieved the status of a family member.

DOG MUSIC

One of the most enduring characteristics of the canid family is the wide range of sounds they use to communicate. Barking, snarling, growling and howling are only half of it. According to David MacDonald and Claude Sillero-Zubiri, co-authors of Oxford University Press's authoritative book *The Biology and Conservation of Wild Canids*, there's a whole lot of talking going on out there in dog land. Along with the New Guinea singing dog, we have squeaking bush dogs, whistling dholes, twittering African wild dogs and 'geckering' foxes (a staccato sound equivalent to the wolf's snarl). Whuffing is also a scientifically acceptable designation for a certain kind of wolf exhalation. 'Huffing and puffing', however, remains in the realm of fairytales.

DINGO OR DOG?

The average dingo is just over a metre long from snout to tip and weighs about 15 kilograms. They are comparable with coyotes in size range, and both species have pricked ears and bushy tails. Although individual dingoes can be seen wandering or scavenging by themselves, they are basically a pack animal. Groups of three to fifteen dingoes will hunt together. Only the pack leaders breed, and the group looks after the resulting cubs. Dingoes aren't fussy eaters, and will consume a very wide range of prey. No doubt this is why they were so successful at spreading across very different terrains. It's also why they are considered one of the culprits behind the disappearance of the mid-range mammals — the notable extinction of a broad range of Australian marsupials that weighed more than a hopping mouse and less than a kangaroo.

The dingo has a coat of few colours: they come in basic black, white, ginger — and possibly, black and tan, although this is subject to some question. Black dingoes are more common in Asia; Australian dingoes are more likely to have reddish-sandy fur, frequently tipped with white on the paws, tail tip and chest. Early explorers did refer to black dingoes in Australia but they are seldom seen here today. Experienced observers can spot an 'impostor' by the presence of a dark stripe along the back, or unevenness in the white patches.

Apart from the fur, one of the most obvious ways to tell the difference between a dingo and a look-alike dog is that, unlike dogs, dingoes only breed once a year. More detailed anatomical examination reveals a narrower-snouted skull, with provision for better hearing (a bigger sounding-box) and bigger teeth.

GIVE A DOG A BAD NAME

Give a dog a bad name, so the old saying goes, and it will behave accordingly. Dozens of popular sayings and slang terms are derived from canid species, many of them related to food or sex, very few of them flattering. It's a dog-eat-dog world!

'Wolf', for instance, is an old-fashioned term for a male who is sexually predatory on females — the term survives today in the expression 'wolf whistle'. The Australian lexical equivalent of 'wolf' is 'dingo'. These days the average 'dingo' is understood to be elderly and fuelled by Viagra. Conversely, young men who prey on older women earn the same title, sans the Viagra. And they might all qualify for the generic slang use of 'dingo' for someone lacking in intelligence. If you 'dingo' someone's food, you pinch it. Not surprising since a 'dingo's breakfast' comprises a yawn, a pee and a quick survey of the territory. Food is conspicuous by its absence.

More recently, the term 'wolf' has been applied to men who are sexually predatory on other men. In Jamaica, though, the 'wolf' is a fake Rastafarian (aka wolf in sheep's clothing). If you describe someone as 'wolfing' their lunch, an Australian or British reader will understand you to mean a person who swallowed their food quickly and in chunks. An American will think the person vomited their meal.

The smaller fox is gender-linked to women, and 'fox' or 'foxy' describes a sexy young female. The more recent Australian variant created by television characters Kath and Kim is the oxymoronic 'foxy moron'. 'Crazy like a fox' expresses reluctant admiration for the fox's cunning which can sometimes manifest as antics designed to distract attention away from her kits.

Alas for the jackals, a beautiful monogamous animal, they eat by scavenging. Thus their name has very negative connotations in English. A human opportunist is often referred to as a 'jackal'. Within the tribe of lawyers, lowly members are frequently asked to scrounge through massive amounts of information looking for tasty legal titbits for their seniors. This activity is known as 'jackalling'.

ARE DINKY-DI DINGOES REALLY MORE DANGEROUS?

Despite their wide distribution throughout South-East Asia, it would appear — at least in the popular press — that only *Australian* dingoes rate as dangerous to people. No doubt this stereotype was given international currency with the release of *A Cry in the Dark*, a film based on the real-life tragedy of infant Azaria Chamberlain, who (as most people now acknowledge) was taken by a wild dingo.

Over the last couple of decades a number of events on Fraser Island, Queensland, have also attracted worldwide attention. In one case in 2004, when an animal entered her family's resort hotel room, a five-year-old girl raised the alarm crying 'Dingo! Dingo!' Fortunately, her parents were in the adjacent room and quickly ran in to find the stout-hearted youngster standing protectively in front of her infant sister. The father managed to chase the animal back outside, and neither child came to harm. A few years earlier, the Gage family were less fortunate. In April 2001, nine-year-old Clinton Gage and his younger brother Dylan were attacked by a wild dingo near their Fraser Island campsite. Despite the best efforts of their father, Clinton was killed and his brother seriously injured. Clinton Gage's death sent shockwaves throughout Australia. Prior to this, most people recognised that a dingo was at least theoretically capable of killing a small child but nobody imagined that a young dingo could kill a nine-year-old and successfully fend off an adult desperately trying to save his sons.

There is enormous controversy about the most appropriate and effective responses to such dreadful events. The subsequent culling of 31 wild dingoes, an animal that is classified as a protected species in Australia, brought into sharp relief a number of issues. Just for a start, the Butchulla people of Fraser Island have a long-standing relationship with the dingo, which is seen as a vital part of their community. They tried unsuccessfully to prevent the cull. When Australian researchers Leah Burns and Peter Howard interviewed a number of people about the Fraser Island dingoes, one of the Aboriginal elders they talked with explained that in her early years it was common for Butchulla women to suckle a dingo pup, which then grew up to guard and protect that particular family.

Another significant issue is that, because Fraser is an island, the local dingoes represent one of the few remaining groups of the pure genetic strain, which is of vital importance if this wild dog is not to be hybridised into extinction. Not to mention the issues as perceived by wildlife managers, tourists and the tourism industry, and the non-indigenous local residents of Fraser Island — all of whom have a variety of often conflicting opinions. Burns and Howard go on to offer a droll account of the different perspectives on the dingoes:

> *It's all pretty straightforward really. The tourists are stupid. The residents short-sighted. The dogs starving. The rangers, who don't know how to look after the island, are over-worked and under-funded. And the government doesn't give a damn ... until somebody dies that is, and then they only give a damn about their political future.*

Such a distillation of complexity could, with a few word substitutions, be applied to a whole range of interspecies conflicts. Of course, Burns and Howard go beyond their brief ironic sketch and reveal the multiple issues involved in dealing with interspecies conflict in a context where interaction with wildlife, in this case dingoes, is a major tourist attraction.

It seems unlikely that the Australian dingo is intrinsically more dangerous than those that live in other parts of the world.

Yet there is no doubt that there has been an increase in dingo aggression on Fraser Island simultaneous with a huge rise in visitor numbers. From the non-indigenous perspective at least, one of the key factors behind dangerous encounters is over-familiarity. When people lose their fear of wildlife, and wild animals through habituation lose their fear of people, the risks multiply enormously for both parties. This can be exacerbated when people feed wildlife. Once humans are a known food source, it's not too much of a stretch for a wild animal to consider humans themselves to be food. After all, the old expression 'don't bite the hand that feeds you' doesn't necessarily make sense from anything other than a human's metaphorical perspective. In the worst-case scenario, such as the Fraser Island tragedy, this includes the loss of human life and the potential loss of whole populations of animals due to retributive culling.

WHEN IS A PARIAH A PURE BREED?

I was once entertained most lavishly at a mansion in Mumbai. Its mews housed gilded Victorian carriages the likes of which I'd never seen outside fairytale illustrations. After dinner, my host offered to show the guests his pariah dogs. I hid my amazement, but wondered what on earth such a man was doing collecting mangy, potentially rabid street dogs.

When two beautiful animals in prime condition bounced into the room, I was stunned to realise they resembled nothing so much as *dingoes*. Thus I learned that the term 'pariah' when applied to a dog can mean one of three things. The term originated in India and in this context refers most correctly to pure-bred Indian wild dogs. However, it is also used for a motley collection of feral dogs of a particularly common type and colour that can be found in many regions of the world from the Middle East to South-East Asia. All of these are likely to be able to trace their ancestry back to the Indian plains wolf. Finally, some 'pariah dogs', including the basenji and the New Guinea singing dog (close cousin of the dingo) are registered dog breeds.

COYOTES

Coyotes can appear similar to wolves, both red and grey, and it's possible to confuse them with the domestic dog. But no one would be likely to mistake a coyote for a dingo, despite their similar size, slender snouts and bushy tails. Not just because they are living on different continents, either. It's the fur that makes the difference, both the darker colour range of most coyotes (rufous through grey), but also the denser winter coat so necessary for the colder American winters.

Like the cowboys who came later, the coyote (*Canis latrans*) was traditionally a creature of the west. Small southern populations existed in Central America, and perhaps even smaller northern ones in Alaska. In fact, most of the original coyote populations were small, and lived in cyclical balance with prey and other predator populations, as well as their Native American neighbours. In sharp contrast to the wolf, however, the coyote was able to take advantage of the post-Colombian population explosion throughout the Americas to massively extend their range. Partly their good fortune was predicated on the ill luck of the wolf. Once numbers of this apex predator severely declined, the coyotes were able to take over the vacant niche. Following the loggers, the coyotes moved further and further east. Whenever ranchers set up for business, coyotes muscled in for the seemingly free feast.

By the mid-twentieth century it was estimated that there were nineteen subspecies of coyotes living in different regions throughout the Americas. Although this is likely to be an exaggerated figure, and taxonomic review is required, it is indisputable that their current range blankets virtually the entirety of Alaska, Canada, the lower 48 states, Mexico and at least the western half of Central America. The coyotes are a canid success story to rival any American dream. And they have also achieved a similar expansion into the broader cultural mythos. To the native people of the Americas, the coyote was always the trickster figure trailing a confusing multitude of contradictory tales. Barry Lopez's marvellous collection of coyote tales in *Giving Birth to Thunder, Sleeping with his Daughter*, celebrates the central place the coyote now occupies in the wider American dreaming.

There is, though, a darker side to this dreaming, one that matches the dingo stories from Australia. In the first years of the 21st century there has been a significant rise in aggressive encounters with coyotes. For a long time, public attitudes to dingoes and coyotes alike tended to run to a rueful, sometimes exasperated, affection. Lock up the lambs, by all means, but nobody thought it necessary to lock up their children. There was no need to be concerned in times of sparser population (both human and coyote). Interspecies aggression was so rare as to be counted absent. Coyotes would only attack humans if they were rabid. That is no longer the case. Beginning in the 1970s, a small number of perfectly healthy coyotes began attacking people.

According to Californian researchers who have documented the increasing incidence of coyote aggression towards humans, the scenario goes something like this. People move into estates and subdivisions in former wild habitat. They landscape. Small wildlife, such as rabbits, rodents and their kin, start to take advantage of the lush new habitat. And where there is small wildlife you can be sure there will be the coyotes that eat them. It may start with rabbits and gophers, but there's little difference between these and small domestic pets from a coyote's perspective. All's fair in love and lunch. And lunch, in the case of coyotes, can be just about anything. They like to dine on fawn and lamb, but rabbits, rodents, cats and dogs will do. And in a pinch they are not averse to a junk food diet of human garbage, both animal and vegetable.

Not surprisingly, they also appreciate a handout — either directly or indirectly. Some people have taken up feeding the coyotes in the mistaken belief that this contributes to the animal's welfare. Quite the opposite is true. The terse expression 'a fed coyote is a dead coyote' sums up the correlation between feeding coyotes and subsequent aggressive behaviour, which in turn frequently leads to the coyotes being classified as a dangerous nuisance and trapped or shot.

Two of the Californian researchers, Baker and Timm, tell a true tale that's redolent of the coyote's trickster persona:

> *At one location in Southern California near the site of a coyote attack, coyotes were relying on a feral cat colony as a food source. Over time, the coyotes killed most of the cats and then continued to eat the cat food placed daily at the colony site by citizens who were maintaining the cat colony.*

This would be funny, except that the results are getting increasingly serious. The following, slightly more ominous, story shows how the coyote got its reputation for being wily. A woman was riding a horse on a trail during the daytime, accompanied by her two big dogs. She spotted a single coyote that immediately took off, one of her dogs in hot pursuit. When the rider and her less excitable dog rounded a corner up ahead, she saw a whole pack of coyotes just behind the dog that had set chase. She concluded it was an ambush and if it were not for her and the horse arriving the outcome may have been quite different.

Timm and Baker and their colleagues list 89 coyote attacks in California, and more than three-quarters of them have happened since the late 1990s. Sometimes people's pets were killed in front of them. It was not uncommon for children to be rescued from potentially fatal attacks by nearby adults. So far there has only been one recorded fatality. In August 1981, a three-year-old girl was killed by a coyote while playing in her front yard. The report noted that the child had bled massively, and that the coyote had broken her neck. The researchers have grave concerns that this singular tragedy could become a commonplace event if steps aren't taken to modify human habits in coyote habitat — and that's just about everywhere. The coyote is nothing if not adaptable. More than 5000 are estimated to live within Los Angeles city limits, that's around eleven coyotes for every 2.5 square kilometres of the city. They have also been sighted within New York City limits. Clearly it's up to us to be even more adaptable and adjust the behaviours that make us vulnerable.

It's not hard to spot the early signs of trouble, say Baker and Timm. It starts at night: coyotes are sighted in the streets and yards of suburbia, they start to approach adults, poach a pet or two. Time passes and the coyotes expand their patrol times to dawn and dusk. Then they grow bolder and wander around in daylight hours hunting pets, eventually chasing joggers, bicyclists, daringly grabbing pets that are leashed. The next and most ominous step is coyotes gathering around playgrounds, parks and schoolyards. Finally, they seem to lose all fear of humans and will attack fully grown adults in the middle of the day. The conclusion is inevitable. When coyotes lose their fear of humans, then humans themselves, particularly small humans, become prey.

People interested in the welfare of both wildlife and children are taking steps to deal with the danger. Re-establishing a healthy degree of respect and wariness on both sides — from human to wildlife and vice versa — seems an essential prerequisite to reducing interspecies conflict.

It's fine for Little Red Riding Hood and her risk-taking behaviour to remain in the realm of fairy tales. It would be a tragedy, however, if wolves and their relatives were driven to join her in the land of make believe.

7

The Woods Today

Bears

Not very many children have their cots stuffed full of soft toy sharks, spiders or wolves. The odd elephant or tiger, maybe. But it's a rare person (in the Western world at least) whose infancy was not graced by at least one teddy bear. Later on, tough adolescents signal their sensitive sides with dashboards full of bears won in skill-tester machines. Whole subcultures of adult teddy-bear lovers exist. They meet up for teddy bear picnics, for heaven's sake.

Bears are reassuring. Bears are comforting. Bears, in at least one instance, are eminently bankable. Think about it. Have you ever heard of someone who was bear phobic? I'm not talking about people who dwell near bear territory and have a healthy respect for their neighbours. I'm talking about the general ether of fear inhaled by people who live in inland deserts and are possessed by a terror of sharks, or people who live in city apartments yet are consumed by a horror of crocodiles. Bears just don't seem to attract that kind of free-floating anxiety like so many other dangerous and deadly animals do. Possibly, it's partly down to the fur. It's a known winner when it comes to capturing human affection.

According to at least one writer, it's not so strange that people have for thousands of years thought about bears as some kind of hirsute human, because bears themselves appear to respond to us as if we are some kind of bald — and frequently badly behaved — bear. Some individuals manage to create and trade on the appearance of having a close rapprochement with bears: the classic example is the well-publicised Timothy Treadwell (see 'Bear films' on page 166), who spent thirteen summers among grizzly bears in Alaska. It's an insider's secret that the particular population of bears Treadwell lived with at Katmai are very tolerant of humans. Wildlife biologist Professor Thomas Smith also worked with the Katmai bears for thirteen years, so he should know.

When it comes to brown bears, though, tolerant and harmless are two different matters. Even experienced professionals can come to grief. Ethologist Vitaly Nikolayenko, for instance, who spent 33 years among the brown bears of Russia. Both Treadwell and Nikolayenko died by bear mauling. Aye, there's the rub, because whatever our romantic longings or professional aspirations might be, humans are not bears, nor bears human.

The blurring of this very basic distinction is dangerous in the extreme: for individuals of both species. People are dreadfully injured and sometimes killed; 'problem' bears are shot. In terms of whole populations, though, there is no doubt that bears are far more threatened by us than vice versa. There are only eight different bear species in the world: brown bears, Asiatic black bears, American black bears, polar bears, spectacled bears, sun bears, sloth bears and the iconic panda. Six of the eight species are threatened or vulnerable, and the dire status of the panda bear is widely known: only around 1200–1600 pandas remain in the tiny bit of remnant habitat left to them. Strenuous efforts both within and beyond China have thus far kept them from taking the final slide into extinction. (These efforts suffered a major setback with the severe earthquake that occurred in Sichuan, China — prime panda habitat — in May 2008. The quake caused the loss of nearly a quarter of a million human lives, as well as widespread damage. The international organisation World Wildlife Fund has 110 panda projects in the area to support its flagship species, and 86 of them were compromised or completely closed down due to the earthquake.)

Clearly bears pose no threat to humans as a *species*. In terms of the danger for individual humans, however, brown bears and black bears are the biggest contenders. Surprisingly, the amiably named sloth bear runs third. This species is actually more of a problem for people than the polar bear, which has had bad press as a fierce predator. Of course, polar bears can and do kill humans, but nowhere near as many as you might expect. In the following pages, we look in more detail at the black, brown and polar bears: where they live, what they eat, how they behave in the wild; and, above all, how we can co-exist with them without coming to grief.

BEAR FILMS

In the first decade of the 21st century two films based more or less on factual human–bear encounters have been released internationally. *Grizzly Man* and *Spirit Bear* provide an interesting study in contrasts, although the central character in each is a bear-obsessed male. The documentary *Grizzly Man*, being directed by the famous German auteur Werner Herzog, is far more cinematically sophisticated than *Spirit Bear*. Yet the latter contains aspects of far greater political sophistication, despite being a romanticised, semi-fictionalised account of real-life activist Simon Jackson.

In *Grizzly Man*, Timothy Treadwell, self-appointed 'protector of the grizzlies' repeatedly states he is on a mission to look after the bears with whom he shares his Alaskan summers. There is no doubting the strength of his conviction and passion yet there is also no evidence that anything Treadwell did had any material and positive impact on bear conservation. The eventual, seemingly inevitable deaths of Treadwell and his partner Amie Huguenard, who were mauled and eaten by the bears, are made even more tragic by this.

Simon Jackson, however, is a different story. As founder of the Spirit Bear Youth Coalition, he mobilised the largest youth environmental movement in the world, made a fundamental contribution to creating a conservation area over two-thirds of the critically endangered spirit bears' habitat in British Columbia, and continues to work towards securing the remaining habitat. Jackson's philosophy and approach is democratic and profoundly engaged with life rather than death.

ASIATIC BLACK BEARS

The Asiatic black bear (*Ursus thibetanus*) is a long-lost relative of the American black bear. The two species are still closely connected phylogenetically (i.e. they belong to the same large related group or phylum), although separated now by oceans and more than 3 million years of evolution. Various subspecies of Asiatic black bears are currently found in two main regions of Asia: south-east from Malaysia through the Himalayas to the Middle East; and in Japan and Korea and west through to Russia. Originally they would have occupied the entire region, but increasing human populations and the resulting fragmentation of habitat mean various subspecies are more and more islanded in particular localities.

Asiatic black bears are marked with a white crescent or v-shape in the fur of their chest, accounting for their other common name of moon bears. It's an appropriate name for a nocturnal animal that is nowadays seldom seen in the wild. For a number of reasons, *Ursus thibetanus* is classified by the International Union on the Conservation of Nature (IUCN) as 'vulnerable'. In Iran and Pakistan, the subspecies *Ursus thibetanus gedrosianus* was thought to be entirely extinct, but one or two sightings confirmed that the subspecies is still hanging on by a thin thread. This recently rediscovered group is classified as 'critically endangered'.

Black bears are tree creatures, and make use of trees for food and shelter. Unfortunately, the Asiatic black bears' predilection for using their sharp claws to strip bark from trees causes conflict when the tree concerned is commercial timber. This activity can result in a death sentence for any local bear, guilty or not. It's a shame because it is simply the bears' nature to use trees for eating and sleeping in. The Asiatic black bear is canny when it comes to denning. Several Russian researchers and their American colleague have been studying Asiatic black bears in Russia. When they compared the holing-up habits of Asiatic black bears with those of the local brown bears, they found most of the black bears used hollow trees, as well as the occasional cave. One wily opportunist had occupied an abandoned brown bear den. (Brown bears put in some hard work hollowing out dens in the hillsides.)

Both American and Asiatic black bears retain the reputation of being gentler than other bear species, though this is less true of the Asiatic black bear. For example, Indian researchers report several case studies of humans being mauled by Himalayan black bears, yet only one such attack proved fatal. When Thakur, Mohan and Sharma reviewed five of these cases they observed some fairly serious head, neck, face and eye injuries, generally from the bear's claws — but the bears never left bite marks or injuries elsewhere on the body. They concluded from this that the attacks were more defensive than aggressive. The bears were not seeking food, and left after rendering the human incapable of chasing them. Sadly for both bears and humans, increasing numbers of humans in the bears' shrinking habitat mean that this kind of attack is likely to increase. I most certainly would not like to be a hapless villager who crossed paths with a nervous black bear. On the other hand, neither would I like to share the fate of many Asiatic black bears that are kept captive and 'farmed' for traditional medicine (see 'Rough trade' on page 178).

The question of how to co-exist with the bears without either party coming to grief is one that has pre-occupied many Japanese people who witnessed the severe decline of the 'Japanese' (Asiatic) black bear when much of their habitat was lost to large-scale tree plantations. Many small regional groups sprang up seeking to prevent the animal's local extinction, and find ways to live together into the future. In 1996, these various groups formed a loose alliance known as the Japanese Bear Network. Networks such as this, with strong regional roots, are one of the most promising ways of working through the problems attendant on conserving species that are known to be potentially dangerous to humans.

DON'T SHOOT THE MESSENGERS

The bears were there first. Twenty-five thousand years later, give or take a century or two, came the people. Those first people have had a long, long time to work out respectful relationships with the prior inhabitants. It's no surprise that the Tlingit and Haida and other native people of Alaska referred to the bears as 'Elder Brother' and 'Old Man with the Claws': to them the bears were spirit messengers, ones who shared their earthly habitat. Not creatures to be treated lightly. Later comers to the region still have a way to go.

In the film *Grizzly Man*, there is an interview with a Parks and Wildlife employee from Kodiak. As a native man, he speaks with the authority of this substantial history of cohabitation. He is reserved but specific in identifying the cause of Treadwell's downfall: lack of respect. Respect in this context is the sum total of thousands of years of knowledge of being with bears. This Kodiak man was simply one of a long line of native people offering the same basic message. While Treadwell may have had good intentions, he lacked such knowledge and did not give the bears the space they needed. In 1912, Allen Hasselborg was mauled by a grizzly in Alaska after ignoring explicit warnings from Tlingit hunter Albert Jackson about the dangers of his reckless and boastful approach to potential bear encounters. He paid a serious price for his hubris but, unlike Treadwell, was lucky enough to survive.

Current wildlife researchers such as Professor Stephen Herrero and his colleague Professor Tom Smith also share this fundamental principle of respect as they go about finding ways for bears and humans to safely share habitat now and into the future.

AMERICAN BLACK BEAR

The American black bear (*Ursus americanus*), unlike the brown bear, is solely endemic to northern America. As with the brown bears, however, the term 'black' is loosely applied to a whole range of colour phases that vary from black right through to white, in the case of the subspecies *U. americanus kermodei*, commonly known as the Kermode bear or spirit bear (the animal Simon Jackson has spent years campaigning on behalf of — see 'Bear films' on page 166). There is even a blue-black variant that lives in north-west British Columbia and in south-east Alaska, known locally as the glacier bear.

On the whole, American black bears are considered less of a threat to humans than grizzlies. By temperament, they are not as aggressive. Their diet is largely, but by no means exclusively, vegetarian; although they will feed on meat opportunistically, they are not in the habit of hunting for it. Their curved claws are designed to help them scale trees, but indisputably the same claws can seriously damage a human being that the bear perceives as a threat.

And just because a bear is considered to be less aggressive, it is certainly not wise to underestimate the degree of danger. Like humans, American black bears are populous and very widespread throughout the United States. Inevitably that means a higher number of human–bear encounters. Those who have had peaceful encounters in the wild are often rapturous, generally mentioning the powerful eye contact. Others, though, have not had such a good experience. And for a small but significant number of people the black bear has been the last sight of their life. A number of newspaper reports of black-bear-caused human fatalities are collected together on the website of the Maine Professional Guides Association.

BROWN BEAR

Brown bear (*Ursus arctos* spp.) is the collective name for a number of different subspecies that live throughout Europe, Asia and the Americas. 'Brown' bears can be various shades of black, red and beige, but the vast majority of them are plain old brown, hence the name. One, *Ursus arctos pruinosus*, rejoices in the common name of the Tibetan blue bear. (*Pruinosus* means 'frosted', so perhaps the poor beast is simply a brown bear that is blue with cold! Not really: such differences can usually be accounted for by adaptation to the colour of the surroundings. Survival does not favour bears that stand out in sharp contrast to their snowy background.)

Different subspecies of brown bear can also vary considerably in size. The Kodiak bear (*Ursus arctos middendorffi*) is almost as big as a polar bear, whereas some European brown bears (*Ursus arctos arctos*) can weigh as little as 45 kilograms. Goldilocks makes a lot more sense when you realise that these relatively harmless creatures dine mainly on plants and small animals. Of course, the Goldilocks story was born of a time when the brown bear was widespread in Europe. These days, the populations are confined to remote mountainous regions in Scandinavia, Romania, Spain, Russia and the former Yugoslavia, where some of them have waxed quite fat indeed, both individually and as a group. (The Romanian dictator Nicolae Ceausescu benefited bears, in one way at least, while being a torment to virtually everybody else in the country. Not that he liked bears or treated them well. He just liked to hunt them. By arrogating the sole right to do so, he created a situation where the bear population of Romania grew apace at the same time as the human population suffered and declined under his rule.)

In certain parts of Russia and Scandinavia individual brown bears grow very large, attaining weights of 500 kilograms or greater. They prefer eating pine nuts to people, but Russia, which has the highest brown bear population numbers in Europe, not surprisingly also has the highest annual death toll from bears, averaging ten people a year. In the rest of Europe, though, fatalities from bear attacks are so rare as to be almost mythical. Almost, but not quite. In 2006, a 42-year-old Finnish man became the first victim of a brown bear attack on record in Finland since at least 1936, if not much longer. While jogging alone during the long summer light of the northern countries, it is surmised he accidentally got between a mother and her cub. Bites from both were found on his body. Being fit, he may have survived the attack were it not that a punctured neck vein allowed an air embolism which was the actual cause of death. He was doubly unlucky, according to De Giorgio and his colleagues, who reported the case in *Forensic Science International*, as embolism is in itself a rare event in animal bite cases.

In the United States and Canada, fatalities from bear attacks are a regular, though not necessarily common, event. Perhaps this is due to a combination of factors. The first is that — at least in Alaska and Canada — there still remain large tracts of the wilderness where bears can live. The second is the nature of the bears themselves. Records of attacks in remote localities may be incomplete, and record keeping has not always been comprehensive, but the rough average is surprisingly similar for brown bears and black bears — around 50 or so human deaths attributed to each over the course of the twentieth century (by comparison only seven or so deaths caused by polar bears were recorded over the same period). At a glance it looks like brown and black bears might be considered equally dangerous, but the numbers only make sense as a percentage of the bear population.

As mentioned above, there are far higher numbers of black bears in North America, and they are spread over a much larger territory than either the brown bear or the polar bear. In other words, they are less aggressive, but you are far more likely to run in to one than you are a brown bear.

WHEN IS A BEAR NOT A BEAR?

Stuffed toy bears were named after American president Theodore 'Teddy' Roosevelt, after he went on a bear hunt. The closest living thing to the toy teddy bear is actually an Australian marsupial — the koala (*Phascolarctos cinereus*), commonly miscalled the 'koala bear'. Koalas are winsome beyond belief to look at, and certainly not likely to kill you. Still, you pick one up at your peril. One Australian politician became a minor cause-célèbre when the koala he was clutching for a photo opportunity urinated on him, providing endless entertainment for cartoonists. The politician should count his lucky stars, though, that the damage was only to his pride: koalas are not bears.

Undoubtedly, the most famous and most dangerous of the brown bears is the awesome grizzly (*Ursus arctos horribilis*). How it came by its scientific name is self-evident, if uncomplimentary for such a beautiful beast. The meaning of the common name is perhaps more ambiguous. The name could easily be (although it is not) a comment on this species' temperament. One study of human–bear encounters in Alaska revealed that grizzlies were responsible for at least 90 per cent of the fatal attacks, and the majority of the injuries. However, the grizzly is so named because the white hairs interspersed among the brown of its coat give it a grizzled appearance. Despite its deserved reputation as a fierce predator, even the grizzly bear is no match for a gun. Like its European cousins, the grizzly is usually found at higher altitudes, although it previously occupied a range of lower territories.

POLAR BEARS

Long, long ago, brown bears wandered into the colder regions of the world, or maybe the home regions of some northern populations became colder around them with the onset of a new ice age. Either way, the species began evolving. Over thousands of years, in response to the new environmental conditions their heads grew smaller and their necks and canine teeth grew longer, until eventually they became the separate species we know today as the polar bear.

Polar bears (*Ursus maritimus*) continue to wander their wild domains, even as those domains shrink rapidly due to global warming. This current change is happening at a rate that far outstrips the ability of most species to adapt to new and frequently adverse conditions. Although there are reports of population increases in specific areas covered by excellent conservation programs, the overall picture for polar bears seems bleak. A 2007 US Geological Survey report estimates less than one-third of the current population of 25,000–30,000 polar bears may survive beyond 2050. In 2008 the polar bears were listed as an endangered species in the United States, which was seen as a key strategy in the struggle to force action to reduce human-accelerated climate change, in addition to protecting the polar bears themselves.

One reason polar bears are a flagship indicator species for global warming is their utter dependence on sea ice, which is melting rapidly if unevenly. Combine this with the fact that they are not a territorial bear and the problem is magnified. Individual polar bears may cover vast areas up to 259,000 square kilometres over their lifetime. Even given that their full lifespan allows them 25 to 30 years for wandering, that's a phenomenal distance. Some of this would be covered on foot, but above all they are sea kings, so quite a lot of mileage can be achieved by 'ice surfing'. Being as much marine as land animals, polar bears are also excellent swimmers, who think nothing of swimming 100 kilometres without pause. Part of the secret is, in fact, their large paws which serve as efficient paddles. Over shorter distances they can achieve speeds up to 10 kilometres an hour.

Given the remote and rugged nature of the polar bears' environment, it is sparsely populated by humans. Perhaps this is one reason that, compared to other bears that are potentially human-killers, they have not caused a huge number of fatalities: seven people in Canada in the last couple of decades; five people in Svalbard, Norway, since 1973; two in Alaska over the last 130 years; and only two in the last couple of centuries in Manitoba, which is, according to one source, the polar bear capital of the planet.

Given their name, it seems straightforward that polar bears' range and distribution is confined to the northern circumpolar regions. Perhaps a more accurate common name, however, would be ice bears or even sea bears, a direct translation of their scientific name. These bears are superbly adapted to their environment of ice and icy ocean in both obvious and subtle ways.

ROUGH TRADE

All bear species on the planet are threatened with the impacts of increased human population, shrinking habitat and climate change. As if this was not enough to have to deal with, some species (including the Asiatic black bear, some Asian brown bears, sun bears and sloth bears) are also subject to farming practices that keep them caged in cruel conditions while being milked for their gall. Added to which, bear paws, skins and gall bladders are sold on black markets for traditional medicine, and sometimes bear meat is sold to restaurants that profit from epicurean thrill-seekers.

It's true that bear gall is known to be pharmacologically effective for certain ailments. According to the World Wildlife Fund, however, viable alternatives exist that are cruelty-free and sustainable. It's a matter of encouraging enough people to use the alternatives rather than drive the bear species to complete extinction.

A WHITER SHADE OF PALE

Clearly polar bears' white fur helps them to blend into the thousand shades of white that paint their polar home. The famous white fur coat is, however, partly an optical illusion. Polar bears have both a guard coat and under fur. The under fur is white, but each of the individual guard hairs that comprise the long layer of fur is actually see-through. Small ears and a small tail, together with the double outer layer of fur and a good inner layer of insulation in the form of 11–12 centimetres of body fat help the polar bears survive in the frigid Arctic.

A plausible-sounding story circulates that says the guard hairs of a polar bear's coat act as a conductor for ultraviolet light, a sort of thermal diode. In other words, they are solar heated! Alas, despite the pseudo-scientific sounding language this is an urban myth. The same applies to the alleged symbiosis between the Arctic fox and the polar bear. It's perhaps less surprising that the more poetic stories of polar bears covering their black noses while they hunt and using hunting tools such as chunks of ice are also part of polar bear mythology. (Scientists suspect the chunks of ice are more likely the result of a giant tantrum in response to an unsuccessful hunt.) And as for polar bears being southpaws — that doesn't even make poetic let alone scientific sense given their northern location!

Adult male polar bears can often reach 3 metres in height standing on their hind legs, and it is not uncommon for them to weigh up to 770 kilograms. One exceptionally massive male topped the scales at just over 1000 kilograms. No wonder they are rated as the world's largest living terrestrial carnivore. The females are about half the size of the males, but that's still substantial. A close encounter with wild polar bears is a scary prospect, even if to human eyes they are a lot more attractive than the former title holder, *Tyrannosaurus rex*.

BEAR BRAINS

A well-known cartoon character claiming to be 'smarter than the average bear' is making quite a boast. Bears, in general, are pretty brainy. It's impossible to make definitive claims about the exact nature of their intelligence, but researchers who have spent enough time observing bears certainly consider them among the most highly intelligent mammals. Comparisons are made with apes, dolphins and three-year-old humans in terms of reasoning power. Their capacity to remember territorial maps far exceeds that of humans, possibly because of the bears' extraordinarily sensitive sense of smell. According to researcher Michael Jamison, the average dog's sense of smell is 100 times better than a human's, and the average hound dog is 300 times better — and the average bear is seven times better than the hound dog. That makes them 2100 times better at discerning smells than us.

MEALS FROM SEALS

The classic image of a polar bear hunting is the long wait by a breathing hole in the ice hoping for home delivery of meals from seals. The local people call these holes *aglus*. Although the seals need to surface every 10 to 15 minutes to breathe, they increase their odds of survival by carving out and keeping open a dozen or more of these holes, so such hunting demands patience. It also takes skill and discernment (see 'Bear brains', opposite). One hapless animal attempted to bite off more than it could chew when the US nuclear submarine *Connecticut* partially surfaced in the polar regions in April 2003. Apparently, the rudder looked like potential food. It was a draw: neither bear nor submarine suffered greatly from the encounter. Although their staple diet is the seal, polar bears will eat kelp and grass when available, and are happy to feed on any dead whale that comes their way.

LIES, DAMN LIES AND STATISTICS

Despite their size, and the rumours that polar bears are fearsome predators and the only bear that will deliberately stalk and kill humans, the statistics just don't bear this out. Among the most reliable keeper of bear attack statistics is Emeritus Professor Stephen Herrero from the University of Calgary, Alberta, Canada. Media frenzies notwithstanding, his calmly compiled data indicates that that there is a steady average of one human fatality per year from polar bear attacks. Herrero endorses the claim that the polar bear is comparable with the black bear in terms of aggression, and both are far more mild-mannered than the grizzly. Neither black bears nor polar bears are inclined to take unnecessary risks: they appear to consider discretion to be the better part of valour. A quick search for information on fatal polar bear attacks reveals that as many, if not more, fatal attacks by polar bears occur in unnatural settings such as zoos.

SHOULD YOU GO DOWN TO THE WOODS TODAY?

Given that the forested areas many bears favour as habitat are shrinking rapidly and given that the same applies to many bear populations, it is worthwhile asking the question: *should you*, in fact, 'go down to the woods today'? The single most important variable influencing human–bear encounters is increasing human population. It's good science, based on decades of research, but it's not rocket science. Bears, on the whole, are remarkably non-aggressive for such large and well-equipped predators; the rare individual black or brown bear is seriously predatory towards humans, but, for the most part, they will not come looking for you.

We do not extend them the same favour. Even when the bear species pose little or no threat to humans, we pose a grave threat to them. And when they do present a possible risk, the odds are still overwhelmingly in favour of humans. In the lower 48 states of America, for instance, the brown bears have disappeared from 99 per cent of their former range (one estimate is that there are fewer than 1000 bears left). Only in Canada and Alaska do fair-sized populations survive. And even then, one study of an Alaskan grizzly population found that the most frequent cause of death for the bears was human-related: 75 per cent for the females and 86 per cent for the males. Whether directly (legal and illegal shooting, car accidents, death by research or relocation) or indirectly (loss of habitat), bears have far more cause to be scared of us than we of them!

For our part, the simplest way of not becoming bear bait is to stay out of their way. Of course, this implies you have a choice, which may not be the case if you are hard scrabbling for survival in the Himalayas, for example. In the United States and Canada, some native peoples have traditionally lived in bear country and continue to do so. In those nations and in Europe, however, many people do choose, for reasons of work or recreation, to spend time in bear country. If all of them took a few simple precautions for staying safe, the likelihood of being killed by a bear would be low. Perhaps, like Australian surfers, the risk could be considered worth it — we cannot reasonably expect wild areas to be 100 per cent safe.

8

BIG CAT MAGIC

Tigers, lions, cougars, leopards and jaguars

IF THEY CARE ONE WAY OR ANOTHER, PEOPLE ARE JUST ABOUT equally divided between those who love and those who loathe the domestic cat. Non-domestic cats are another matter. All the predators that eat humans inspire terror, by any measure a reasonable response to something that can kill you. Yet only the wild cats also inspire almost universal respect and frequently deep affection. Nobody ever attached the word 'noble' to a shark or crocodile. They lack the beautiful pelt, the lustrous eye, the aristocratic tilt of the head that attracts much more than it repels. We are obsessed with cats in the wild, in zoos, in film and book and song, in dream and in reality. The ancient Egyptians are not alone in their worship of cats; but there is a dark side to this worship.

It's a very small step from worshipping the mystique and power of these animals to desiring to possess such power for ourselves. Sympathetic magic is one of the major drivers of the deadly black market for animal parts. *Maybe this tooth, this claw, this skin will transfer the animal's power to me*. Along with habitat loss, the illegal trade in poached animals means we are losing far too many of them. In the case of the wild cats, there are 36 species of big and small wild cats that belong to the family Felidae. Of these, seventeen are threatened and five endangered or critically endangered according to IUCN criteria. In response to effective conservation measures a few populations of wild cat species are thriving, but most are not. This puts us in a strange position. We have been pitted against these predators in a life and death struggle since time immemorial, and until very recent historical times you could say that neither side possessed the capacity to vanquish the other. Hunger is a great motivator. Not only do wild cats eat humans, but all the big carnivores, including humans, have a penchant for pinching each other's kills, given half a chance. One study in Uganda reported that it is still not uncommon for rural people without firearms to try scavenging from lion and leopard kills. In the reported cases, sometimes the wild cat was driven off, sometimes it was killed, and sometimes the humans lost their lives in the process.

In terms of ability to kill humans, five felid species top the list. Four of them are big cats that belong to the genus *Panthera*, in the subfamily Pantherinae: in order of danger to humans, these are tigers, lions, leopards and jaguars. These could be dubbed the 'roaring fours' as the ability to roar is one of the distinguishing features of the genus *Panthera*, the lion being the loudest. The sole so-called small cat to pose a significant threat to people is the cougar (an animal of many names) that is classified as genus *Puma*, in the subfamily Felinae. Just because they can and do kill people, however, does not mean they are necessarily doing so for food. Other motives can be defence of territory and protection of young. Tigers, in particular, are frequently known to attack and kill people and simply leave the body alone. Although the phrase 'man-eating' (fill in the animal and species of your choice) is commonplace in sensationalist media reports, it's a mistake to think that all lions or tigers fit the category. It is also a mistake to think there is no such thing. A certain small percentage of these animals do earn such a title by repeatedly attacking and killing humans for food. 'Man-eating' lions may be the very stuff of nightmares and horror films, but they are real.

THE TSAVO LIONS

Two of the most notorious of these made headlines around the world in the late nineteenth century. This pair plagued the builders of the Tsavo Railroad Bridge in Kenya. What a thankless task those builders had — working long hours in rough conditions, and subject to nightly death threats by the local carnivores. Over nine months in 1898, an extraordinary number of workers were killed by the maneless Tsavo lions: some claim up to 140 in total. Several reasons have been advanced to account for such persistent preying on humans. An outbreak of rinderpest had greatly reduced the lions' normal prey. Post-mortem dental examination of the lions revealed disease that may have reduced their normally phenomenal jaw strength, and consequently their ability to prey on huge herbivores. A more macabre hypothesis is that workers who had died from other causes were not given adequate burial, and thus the lions became habituated to the taste of human flesh.

The man who shot the Tsavo lions, Lieutenant Colonel John Henry Patterson, was the type model of the 'Great White Hunter'. He made a tidy sum by writing a book detailing his exploits, and then topped up the profits by selling the lions' skins to the Chicago Field Museum for $US5000, which in 1924 was a small fortune. It's possible he had slightly inflated the numbers of people these lions had killed, but the skins bore out his testimony that both of the Tsavo lions were close to 3 metres long from muzzle to tail tip: as big as a tiger! The audience for such stories constantly renews itself and in the 1990s, almost a century after the attacks, a movie was released based on the events, *The Ghost and the Darkness*.

Seen from the distance of more than a century, such stories do seem to attain the status of tall tales and legends. Still, such seemingly mythical man-eaters have not quite vanished into history. In the early 1990s the villagers of Mfuwe in Zambia were beleaguered by another huge maneless lion that killed six people. By all accounts, this lion also dragged a bag of clothes belonging to one of the victims through the village to the river, where it continued to bat it about in play. Anybody who has witnessed a domestic cat toying with a mouse could find it in their hearts to sympathise with the locals, who witnessed this and came to the conclusion that it was a matter of sorcery and evil.

After several groups tried and failed to locate the right animal, the 'man-eater of Mfuwe' was eventually shot by Chicago conservationist–hunter Wayne Hosek, much to the relief of the locals, including the remaining lionesses: six of their sisters had been shot in the mistaken belief that they might have been the perpetrator. This hunting was auspiced under the Luangwa Integrated Resource Development Project, which has had remarkable success in reducing poaching and enhancing wildlife conservation at the same time as improving economic and social conditions for local people. In terms of dealing with interspecies conflict, this approach of controlled, sanctioned shooting may be the way of the future. (See 'Can you have your lamb and lion too?', opposite.)

Deliberate predation on humans and death tallies on such a scale are absolutely the exception, yet clearly the people of Mozambique, Tanzania, Uganda, Zambia and Ethiopia all have reasonable grounds to fear for their lives and livelihoods due to the depredations of lions. Just as the Bhutanese, Bangladeshi,

Indian, Burmese and Nepalese people who live near the Sundarbans National Park have reasonable grounds to fear the Bengal tigers that dwell there.

It's the same problem, over and over again. Predator species require prey species and large territories. Humans herd prey species and require even larger territories. Inevitably there is conflict. And while individual humans may suffer genuine loss, sometimes catastrophic loss, the predators are, overall, even worse off. All of the dangerous wild cat species are compromised, and the tiger is critically endangered in much of its former range. Driven by the demands of our most primitive fears, we look set to win the most pyrrhic of all victories against the wild cats. Before it's entirely too late, it might be worth asking ourselves: even if they can kill us, do we really want to live on a planet without them? Some useful answers may come from researchers working in the relatively new field of interspecies conflict.

CAN YOU HAVE YOUR LAMB AND LION TOO?

Ecology is a difficult game. It's not always easy to intervene in complex systems and have uniformly positive results. Until recently humans were, for the most part, quite sanguine about the elimination of other species with whom they found it difficult to co-exist. This is hardly surprising given our habit of attempting to eliminate whole groups of other humans with whom we find it difficult to co-exist. The general approach to dangerous and difficult animals was Dalekesque: exterminate!

exterminate! Bounties were placed on an animal's head if it so much as poked its nose up within a 100-kilometre radius of domestic livestock. More than one species was hunted to extinction with nary a qualm.

In some quarters, however, people started having second and third thoughts when this slide to extinction began resembling a landslide. A sea change in public policy in various countries has had some dramatic results. Protection and re-introduction of large predator species into parts of their former range occasioned mixed feelings. Conservation success stories often bring us back to square one: quite a lot of people are not keen on sharing with bears, wolves, cougars, tigers or lions.

A whole new field of research has arisen to examine ways to confront these issues dispassionately and fairly, and devise strategies that will allow ongoing co-existence and minimise conflict. Different ways of managing traditional farming, different types of farming, tourism, even limited forms of hunting may end up being part of the mosaic.

TIGER, TIGER, BURNING LOW

'Tyger, tyger, burning bright in the forests of the night': Blake's immortal words distil the experience of thousands of years of human co-existence with tigers. However uneasy, however endangered we felt, we never dreamt the tigers were not immortal. With good reason, Koreans acknowledge the tiger rather than the lion as the 'King of Beasts'. The tiger is after all the largest of the cat family. Males are bigger and longer than females, averaging around 3 metres in length from muzzle to tail tip. The average male tiger weighs between 190–200 kilograms, but one formidable beast was recorded at 320 kilograms — three times the weight of a large human male.

Sadly, the tigers' light burns very low today. And there are so few forests left. Tigers (*Panthera tigris*) are strictly Asian animals and the common names of eight alleged subspecies sound like a roll call of places where they used to thrive in the wild: Balinese, Sumatran, Javanese, Indochinese, South Chinese, Bengal, Siberian and Caspian tigers. Little more than a hundred years ago there were up to 100,000 tigers spread out across the whole of this extensive range. Now, at least three of the subspecies are considered extinct: the Caspian tiger (*Panthera tigris virgata*), Balinese tiger (*Panthera tigris balica*) and Javanese tiger (*Panthera tigris sondaica*). Others are nearing extinction. The South China tigers (*Panthera tigris amoyensis*) have fewer than 30 or 40 individuals left in the wild. Forty years ago they numbered in the thousands, but during the reign of Mao Zedong a bounty was put on their head and the stark result is almost certain extinction.

In sharp contrast, the Indian government under the leadership of Indira Gandhi responded to recognition of tigers' endangerment by initiating large scale, ongoing conservation efforts in an attempt to secure the future of their very own Bengal tiger (*Panthera tigris tigris*) — the 'tiger's tiger'. Starting in 1973, 'Project Tiger' created 40 tiger reserves across a range of habitats.

This effort was massive, and at first it did have a major positive impact on the tiger population. For a while, the Bengal tiger population seemed to be holding steady and even increasing. There is a wide margin for error, due to disputes about the method of calculation, but it was estimated that largely due to these efforts 3000 to 4000 Bengal tigers still roamed the wilderness between India, Bangladesh, Nepal, Burma and Bhutan at the turn of the century. Unfortunately, more accurate census data from 2008 indicates a far lower figure of between 1300 and 1500 wild tigers left. A depressingly low number, as such small populations are at a much increased risk of decline and extinction. Most other species of tiger have far lower population numbers, so the overall picture for tigers is grim.

Tigers are the most dangerous wild cats for humans. Statistics are not globally comprehensive, but average numbers of fatalities in tiger 'hotspots' serve to illustrate. In certain parts of India — Madhya Pradesh, for example — tigers take a small but significant annual toll of a dozen or so people. The highest numbers of tiger attack victims live in or next to the Sundarbans, which lie between India and Bangladesh. The Sundarbans is tiger central, one of the few regions of extensive wild habitat left for the Royal Bengal tiger. An average of 100 people a year are killed, but it's not 'equal opportunity' killing. Experts note that only around 30 per cent of Sundarban tigers are actively aggressive towards humans, and only 10 per cent of the aggressive tigers qualify as dedicated people-eaters. Most tigers much prefer to avoid humans; it's just getting harder and harder for them to do so.

THE FRANKENSTEIN TIGER

When you arrive in a completely new world, it offers some comfort to name it with familiar names. Australia has the koala, which is not a bear. And the striped thylacine or Tasmanian tiger, which was no tiger, Tasmanian or otherwise. Alas for the thylacine. We made it a monster, a Frankenstein, then hunted it to death. It was never known to kill a person. It most surely did not kill all the livestock for which it stood accused and was executed.

POACHERS PREY ON TIGERS

Judged in the light of this, the underpaid and underequipped guards who continue to devote their lives to tiger protection must have extraordinary courage: both physical and moral. At least 50 guards lose their lives to poachers each year, and twice that many are injured. These guards have phenomenal knowledge of the animals they are protecting, understanding their habits in health and illness, noting their signs of distress. 'The animals recognise us and stop when we whistle,' said Bisht who has patrolled the Himalayan jungles for close to 30 years in defence of the tigers and other protected animals. People with his kind of knowledge are almost as endangered as the tigers themselves. Most are in their fifties or older, and there is no incentive for new workers to replace them. The pay and conditions are very poor. Bisht, along with many of his fellow workers, is respectfully cautious of the animals he is protecting, but justifiably lives in far greater fear of the poachers that stalk the animals.

In countries where there are radical extremes of both wealth and poverty, the huge profit margins from poaching will always be a temptation. And extreme inequity is always accompanied by its shadow: corruption. Poachers have nothing to lose by being ruthless about life, either human or other animal. Tiger teeth, claws, skins, live cubs can all be sold on the black market. The animal itself is so mysteriously powerful and beautiful that long ago people concluded that they would very much like to partake in that power. It was hoped that by owning various parts of a tiger there would be a magical transference of the animal's essence. In some recent instances, though, this hideous but thorough economy has escalated into extreme waste. In certain areas, poachers have been known to abandon tiger skins in the forest because the bones were easier to conceal and could fetch more instant cash, with less trouble than hiding and selling off the skins. One tiger reserve in Rajasthan lost all eighteen of its tigers to poachers over a six-month period in 2004, due largely to corruption and mismanagement. Even zoos offer no guarantee of sanctuary. In 2000 a one-year-old female tiger was killed and skinned at a zoo in Hyderabad. Four zoo employees were suspended, and more were implicated in the crime.

Indian conservationist and naturalist Valmik Thapar has long been an active critic of such corruption and the toll it takes even on the veracity of scientific reports. Since succumbing to the spell of tigers in the 1970s, Thapar has tried every means in his considerable power to fight for their continued existence. Time and time again he has challenged the systematic corruption that augurs so ill for the tigers' continued existence. He is not optimistic about their long-term survival. If tigers do hold on through the 21st century it will be due to people such as Thapar, and many other wildlife biologists who work ceaselessly to this end. It will also be due, in no small part, to the frontline workers who continue to risk so much for so little material reward.

WONDERING WHERE THE LIONS ARE?

Unlike the tigers, which never set paw in Africa, one species of lion dwells in Africa while another subspecies remains in Asia. Compared with tigers, the African lion (*Panthera leo*) population is in a less precarious position. Still, the same story of falling population applies. Conservative estimates of African lion numbers in the first decade of this century are between 16,500 and 30,000, which represents a drop of between one-third and two-thirds of the population alive at the start of the 1990s. This is considered a sufficiently serious problem for the African lion to be classified as 'vulnerable' according to the IUCN classification scheme.

Far worse off, however, is the Asian lion (*Panthera leo persica*), a remnant population from the days when the lion roamed a much wider territory across Africa, Eurasia and even North America. Slightly smaller than the African lion, and with a scruffier mane, the

Asiatic lion is definitely down on its luck. Only 300 or so of them are left in the wild, all of which can be found in the Gir forest of Gujarat in India. Predictably, interspecies conflict is a major problem. Human fatalities and livestock losses are ongoing problems. For instance, from 1978 to 1991 an average of two people each year were killed by Gir lions. Nonetheless, the Asiatic lion is less of a danger than its African counterpart on an attack-by-attack basis. One study found that only 14.5 per cent of attacks had fatal outcomes.

In contrast, the African lions are significantly more successful at the killing game. The lion is a superbly equipped predator. Light-boned and muscular, it embodies the perfect balance of opposite forces. The lion's heavy forepaws can break a spine with one blow; the pads in that forepaw are very fleshy so it can tiptoe (literally — the lion is a digitigrade walker) quietly up to its prey. Hunting is a precisely orchestrated activity, most efficiently carried out by lionesses, which are lighter and less hampered by the big hairdo of the male. Once satiated, the lion rests in the lordly manner that earned it the title King of the Beasts everywhere in the world bar Korea. A pride can rest on their hunting laurels for four or five days before needing to hunt again. Each lion may account for the death of fifteen to twenty large herbivores a year, which is not enough to make much of a dent in the population of these species.

Very few lions have humans on their list of species killed. A most comprehensive survey of those that do was carried out by researchers Treves and Naughton-Treves in Uganda. They substantiated records of 275 attacks over a 72-year period. Only

25 per cent of the victims survived. At least one-third of the 175 reported fatalities over a fifteen-year period from 1989 were in Tanzania. Another pattern also emerged from their study. Males were more likely to experience being attacked by a lion (given that their daily lives require them to be out hunting themselves); males were also more likely to survive such an attack. Attacks on women and children, though less frequent, were almost inevitably fatal.

Despite its fearsome capacities as a hunter, the lion is not immune to multiple threats to its existence. It leads a relatively short life in the wild (averaging twelve years in comparison with sixteen or so years in good zoo conditions). Apart from poaching and habitat loss, the lion population is declining dramatically due to diseases, including canine distemper, tuberculosis, feline immunodeficiency virus (FIV) and HIV.

COUGARS: THE AKA ANIMAL

Cougars (*Puma concolor*) are one of two home-grown wild cats of the Americas, their more southerly counterpart being the jaguar. Although cougars are lighter in weight than jaguars, they still rank as the fourth heaviest feline species in the world. Nonetheless, scientifically speaking they are classed as a small cat. They have had more than one scientific name, formerly being classified as *Felis concolor*, but that's nothing compared to the number of common names. Cougars aka pumas aka panthers aka mountain lions aka catamounts aka painters aka mountain screamers. These animals have far more names than the domestic cat has fabled lives: 40 different names in English alone, which is ample evidence of just how big their range was in recent historical times. New neighbourhood, new name.

Like all large predators, the cougars suffered population drops and reduction of range as the human population expanded throughout the United States and Canada during the nineteenth and twentieth centuries. By the close of that century, though, wildlife policies had become much more conservation-oriented, and the cougar was subject to a shift in status, from a pest species with a bounty on its head to one that can only be hunted under restricted conditions. In California, it is a protected species. The resulting upswing in the cougar population and their return to some of their former haunts has, predictably, received mixed responses from the public. On the one hand, it is counted as one of the more cheerful conservation stories. In an era of severe and multiple environmental threats, it is heartening to many to hear some good news.

On the other hand, cougars can and do kill people, most especially children. The death of a child is a tragic and emotional topic, and one that is difficult to place in perspective. Each year 50,000 children die in the United States, according to *The National Centre on Child Fatality Review*. Although cancer is the leading cause of disease death among children, at least 2000 children die *annually* from abuse, or to use the chilling Australian term 'non-accidental injury'. Modern Western cultures tend to insulate against death in all its guises, and death by wildlife seems shockingly primitive and somehow wrong. It is also, to place it in perspective, extraordinarily rare. The animal most likely to kill a child is an adult human.

Indisputably, though, the numbers of fatal cougar attacks have risen since the 1990s. Wildlife ecologist, Professor Paul Beier made it his business to find out exactly how many people had been killed by cougars in North America, including Canada, over a 100-year period starting from the last decade of the nineteenth century. Most of the time, people can walk right by concealed cougars and not even suspect their presence. Sometimes, hikers or campers will sight a cougar. On even rarer occasions, the cougar will attack a human outdoors, sometimes even near dwellings and in residential areas. But in 1951 a telephone linesman inside his remote cabin must have really drawn the short straw statistically when a cougar actually leapt through a closed window to attack him. He survived the experience, as do most adults who have run-ins with cougars.

Beier's work uncovered the fact that the sum total of verifiable fatalities over a 100-year period was *eleven*; all bar two were children. One eighteen-year-old, Scott Lancaster, was killed while jogging alone on a trail near his high school in Colorado on 14 January 1991. (Lancaster's unlucky and untimely demise had a fictional reprise as one of the more unusual death scenes to open an episode of the television cult classic *Six Feet Under*.) The other adult death was part of a double fatality caused by rabies acquired during the attack. More than half of all the fatal attacks have occurred since the 1970s.

Unless the cougar is rabid, however, it will almost certainly back off in the face of strong opposition. Plenty of people have chased them off with sticks or stones. And some have managed this just by standing their ground and shouting. It's a shame Chicago officials didn't have this information to hand (or at the very least a tranquilliser gun) when they confronted a cougar wandering the streets of their city in April 2008. In what could be seen as a rather drastic response, the animal was gunned down in a hail of bullets. Shades of John Dillinger! Must be the way they still do business in Chicago. It's remotely possible that the unfortunate animal may have travelled some thousand kilometres from its natural range in search of a mate, but it is far more likely to have been a pet that got too hard to handle, and either escaped its handler or was dumped. Either way it certainly had to be removed from city streets near a school, but was it really necessary that the removal method be so permanent? (In stark contrast, a month later Malaysian Department of Wildlife and National Parks officers went to Segamat, Johor, and managed to relocate two rogue elephants that had been terrorising local villagers using the simple expedient of a tranquilliser gun.)

PANTHER UNRAVELLING

Panthers are black, right? Except when they're an endangered Florida cougar, commonly known as the Florida panther, which has fur the colour of tawny sand. Then again, the normally orange-tawny jaguars that live in the southernmost part of the

United States right through Central and South America from time to time sport a melanistic form. It goes by the name of black panther. The even rarer albino jaguar is (you guessed it) the white panther. Just to round things out, the jaguars' Old World cousins, the leopards, also, though less frequently, produce a fully black form. These, too, are called black panthers. In no case is the common name panther illuminating about the taxonomic relationships among the animals to which it is attached. Now you know why scientific names are so handy. Not that they are immutable or free of confusion, but at least they are a bit more systematic.

LEOPARDS AND JAGUARS

Even combined, the other pair of big cats to make up the 'roaring fours' barely match the danger of tigers, lions and cougars. True, some leopards (*Panthera pardus*) have been documented 'man-eaters'. The Panar leopard, for instance, allegedly killed at least 400 people in the Kumaon district of northern India in the early years of the twentieth century. Ironically, it only began its career after being injured by a poacher and finding itself unable to hunt its normal prey, or so the story goes. By contrast the Rudraprayag leopard that terrorised the people and pilgrims of Uttarakhand in central India only took 125 people. But this was a decade or so later, and perhaps the figures are more accurate. Both leopards were shot on commission by the legendary Jim Corbett (see the

Damascus Awards on page 206). Being night hunters, leopards have even been known to sneak indoors and attack sleeping people. These man-eaters are definitely the exception, however. According to J.C. Daniels the number of leopard attacks in India has declined significantly since the beginning of the twentieth century. For their part, leopard populations appear to be holding up better than those of the other big cats, but a number of subspecies are seriously threatened.

The enigmatic jaguar (*Panthera onca*) has never been classed as a dedicated man-eater. Hunted out of most of its former range in South America, extinct in Uruguay and El Salvador, gone from the southern United States and northern Mexico, the jaguar's last refuge is the remaining rainforest of the Amazon basin. On rare occasions the jaguar does kill a human, but it has not been known to do so 'unprovoked'. As most of its victims tend to be children, it's important to be clear that 'provocation' is not necessarily deliberate or aggressive. Rather, the term applies when something appears a provocation from the animal's point of view, even if it would not necessarily seem so to a human witness. Although the wild jaguar does not seem to be the most dangerous of the big cats, the jaguar at Melbourne Zoo kept the keepers and the zoo veterinary on their toes by launching itself at the back door of its cage every time they passed by. Perhaps it was just that particular animal, or perhaps the jaguar is less able to adapt to a life in confinement.

REVENGE OF THE POST-COLONIAL CARNIVORES: A BRIEF HISTORY OF BIG CATS ABROAD

There is something strangely compelling in the brief history of the big cats of the so-called New World. Fantastic rumours must have reached Europe from the first travellers to reach the shores of Africa and Asia and return. Talk of strange, terrifying creatures straight from myth or story or some vague race memory of earlier creatures such as the cave lion (*Felis spelea*), which dwelt in the British Isles during the Neanderthal era. During the long reign of European colonialism patriots and expatriates talked of giant cats that could devour humans. Such talk occasioned barely concealed shudders of visceral delight (the seventeenth-century thrill equivalent to watching horror movies). During the eighteenth and nineteenth centuries, the grandiose figure of the Great White Hunter stalked across the Western imagination, and there was much slaughter and trophy collecting. Throughout the twentieth century his day gradually waned, and the big cats were re-cast as objects of conservation. But the lust to collect was simply transferred to live animals.

Now in Europe, in Britain and most of all in the United States there are any number of big cat owners, very few of whom know how to deal with such an animal once it passes from cute cub to full-grown carnivore. Meanwhile, the cats become habituated to humans and lose all fear of them. Inevitably, people then rediscover the fact that these animals can and do kill humans. Even the incomplete number of exotic big cat attacks each year in North America is startling. Since 1990, at least 100 people have been injured by lions and tigers, twelve of them fatally. Compare this with the number of people (ten or twelve) killed by the wild and non-exotic cougar during the entire twentieth century.

Numbers don't quite convey the human cost. At seventeen, Haley Hilderbrand was poised on the edge of adulthood, her whole life to come. A decision to have her senior photograph taken with two cute tiger cubs took that future from her. Instead of cubs, for reasons unknown, the facility brought out a very large Siberian tiger on a simple leash, with fatal results. After Haley was attacked and killed, the tiger was shot. Haley will not be forgotten, though, because her name is attached to the proposed US federal legislation to protect against such senseless tragedies by prohibiting direct contact between visitors and animals. Although several American states have already enacted such bans to good effect, states such as Florida continue to be highly unregulated and consequently to have correspondingly high numbers of big cat 'incidents'.

EXACTLY WHO IS RED IN TOOTH AND CLAW?

The world's most endangered big cat is so rare that most people have never even heard of it. The amur, or far eastern leopard (*Panthera pardus orientalis*), lives in the remote region between

Russia and China, where two, at the most three, dozen of the animals survive in the wild. Amurs used to roam all through north-east China and the Korean peninsula, but now they only linger in this last borderland. And for all the wildlife biologists, conservationists and animal lovers who desperately want them to survive, there are others who do not scruple to kill and further drive the species towards extinction. In 2007 one of only eight healthy adult female amurs was found shot and bludgeoned. She was the third such animal to be killed in the previous five years, most likely by poachers. The struggle for a species' continued existence does not get much more raw and brutal than this. Part of the problem lies with the sheer amount of high-quality habitat (well stocked with prey species such as deer) that is needed to support key predators such as leopards. According to Dr Dmitry Pikunov of the Russian Academy of Science, each adult male amur requires around 500 square kilometres of territory (that's over half the size of New York, which is around 800 or so square kilometres), a range which would also support a couple of breeding females and their cubs. To date, the amurs are still breeding, but unless there is significant ongoing habitat protection it is only a matter of time before this beautiful animal disappears altogether.

THE DAMASCUS AWARDS

Perhaps the love–hate ratio has always been more finely balanced for the big cat species. Certainly it seems almost commonplace for former big cat hunters to turn into conservationists. Perhaps they are all just following the shining example of one of the earliest and most famous, Jim Corbett. Born in India of Irish ancestry, Corbett is no relation to the heavyweight boxer of the same name, who was more widely known as 'Gentleman Jim'. Yet Edward Jim Corbett most certainly deserved that title, too. In conservation circles, he was without doubt a heavyweight.

At the age of ten he began his hunting career by shooting a leopard. In the ensuing years he developed a great proficiency at hunting, being by all accounts both brave and skilful. In later years, he preferred to operate alone, although he spent some time as a hunting guide. It was this occupation that eventually shifted his perspective on hunting. The senseless massacre of more than 300 waterfowl by a party he was leading made him vow to give up the so-called 'sport of kings'. His youthful passion for hunting was superseded by a love of photography and a dedication to the conservation of wildlife and the habitat needed to sustain them.

Throughout his life, though, Corbett was always willing to use his hunting skills when big cats posed a serious threat to human life or livelihood. In this capacity he was responsible for hunting and killing several animals, including the Champawat tigress and the Mohan man-eater in India. These animals are alleged to have killed extraordinary numbers of humans, sometimes numbering more than 1000. Although these wild figures are unlikely to hold up under scientific scrutiny, it remains true that the animals he shot later in life were definitely repeat killers, mainly of women and children. For this reason he was held in very high regard by the local people, some of whom saw him as a *sadhu*, a holy man.

No one knows what the cats think, but Corbett did much to protect them, too. He founded the All-India Conference for the Preservation of Wildlife. He was responsible for securing the first, and (after additions of land) the largest, national park in India. Corbett National Park in Uttaranchal, India, was renamed in his honour in 1957. Later in his life he lived and worked actively for conservation in Kenya. Altogether, Corbett is not only the archetypal recipient of a Damascus Award, he can be credited with virtually inventing the genre.

9

The Deadly Vegetarians

Hippopotamuses, rhinoceroses, buffalo, bison, moose and elephants

Vegetarians are generally considered to be peaceful, passive beings, untroubled by the blood lust rightly or wrongly attributed to the carnivore. For this reason, it is generally carnivores that spring to mind when the terms 'dangerous' and 'animal' are spoken together: sharks, lions, tigers, crocodiles — all the toothy denizens of our nightmares that have been written about elsewhere in this book. However, as anyone knows who has had cause to go around the long way to avoid the bull's paddock, all those grass-chewing bovines aren't nearly as safely bucolic as they are made out to be. And people who share territory with some of the mega-herbivores are well aware that blood lusts are not all attributable to the desire for meat. When the hormones drive, even generally laid-back animals become extremely fierce. Anyone unfortunate enough to encounter an elephant in 'must' knows exactly how dangerous a herbivore can be. The two relevant factors are size and unpredictability. No one has ever been preyed on by an elephant or a hippopotamus, because they don't eat prey. But plenty of people have been charged, trampled and gored by them, frequently fatally. So 'follow me, follow, down to the hollow', and there let us meet the first of our deadly vegetarians.

HIPPOPOTAMUSES

In their old song, Flanders and Swann might have painted a jolly picture of the hippopotamus wallowing in 'glorious mud', but they obviously did so from the safe distance of Britain. If they were resident in, say, Uganda or Malawi or Kenya, they might have had an altogether different impression of this mega-herbivore. Just ask researchers Adrian Treves and Lisa Naughton-Treves. They combed through the diligently kept records of the Ugandan Game Department from 1923 to 1994 looking for accounts of carnivore attacks on humans. Along the way they came across records of human casualties, some fatal, caused by an assortment of other animals including buffalo, elephants, baboons and chimpanzees. When they crunched the numbers, it turned out that the hippopotamus (*Hippopotamus amphibius*) was responsible for more fatalities per individual attack than any other animal, including its closest rival — the lion. On the whole, the researchers felt that the high number of deaths most likely reflected the fact that fishers attacked by a hippopotamus face the secondary but very real danger of death by drowning.

In fact, that's one of the significant difficulties facing humans who wish to avoid the hippopotamus. As a tourist, you do get a choice of whether or not to spend time in the water where hippopotamuses spend their days, sometimes completely submerged, sometimes with only their eyes and nostrils visible above the surface. As a local dependent on fishing for your livelihood, you have little option but to take the risk. No wonder so many African fishers rely on that old standby, the hat with eyes on the back. This ploy is used in various countries to ward off dangerous and nuisance encounters with animals, from pouncing tigers to swooping magpies. Animals are wired to pay attention to other animals' eyes — it's a survival mechanism. They are far less likely to attack an animal they think is looking straight at them.

OF LANGUAGE AND MUSIC

The word hippopotamus is from the Latin and means 'river horse', as most trivia and quiz fans know. It's less widely known that in terms of taxonomic affinities it might better be named *terracetos* — pig Latin for 'land whale', as the whale is one of its closer relatives. And though I've yet to see a new age recording of 'hippopotamus music' they are bilingual in the sense that they make low frequency communicative noises both in the air and under water — and often in both mediums simultaneously. This makes a lot of sense for an animal that spends up to 90 per cent of its life in water. The aptly named William Barklow, who made a study of hippopotamus acoustics, admits that it's unclear as yet what the subaqueous sounds mean, but they produce fountains of water similar to the bubble blasts of grey whales. Barklow hypothesises that these 'may facilitate long-distance "chain chorusing"'. So perhaps the hippopotamuses are more akin to the jolly wallowers of the Flanders and Swann song than I first thought. Or, on the other hand, they might be chorusing some militant hippopotamus hymn which would be most helpful to know about so we could get out of their way! Further research is needed.

The hippopotamus and rhinoceros are pretty much equal runners-up in the world's largest living land animals, with the title held by the elephant. At around 1.4 metres tall at the shoulder, hippos are shorter than the average adult human, but their length (3.3–3.45 metres) is almost twice the height of an average adult human. And — despite the growing obesity epidemic in the developed world — none of us are a match for the adult male hippopotamus, which weighs around 1.6–3.2 tonnes (the upper limit is almost exactly the weight of a modern LandCruiser). That's impressive, yet the most awesome statistic of all is the animal's jaw span. When it gapes, the hippopotamus's jaw can open to a phenomenal 150 degrees (in comparison, even the most extravagantly yawning human can only manage a measly 45 degrees), displaying a pink mouth and canine teeth as long as 30 centimetres. (Hippopotamus canines are currently a legal substitute for elephant ivory!) This is no mere yawn of boredom, and quite the opposite of a welcoming smile. As any hippopotamus could tell you — a yawn is a major threat display. It is one of the commonest signs of aggression, along with the charming male habit of spraying his opponent with dung.

Despite their bulky bodies and squat legs, hippos are surprisingly fast on land (speeds of 20 kilometres per hour are regularly achieved over short distances). It is their habit to emerge on to land at night to graze. But when threatened hippopotamuses will head straight for the nearest waterhole; any humans in their direct line of flight are at grave risk of being trampled. And, although it is commonly rumoured that the hippopotamus can only run in a straight line, other sources attest that it can 'turn on a dime'.

Unlike for many other African animals, there is not a great deal of scientific research available on the hippopotamus. This lack is being slowly rectified by the Hippopotamus Specialist Group of the IUCN, among other groups. One fact is clear, according to the IUCN: the hippopotamus may be huge and dangerous, but it is classified as vulnerable, with less than 150,000 left in the 29 African countries it calls home.

HIPPOPOTAMUS HOAX?

The article is entitled 'An Unusual Housemate', and it is accompanied by an interior shot of a seemingly contented baby hippopotamus fronting up to a kitchen bench for lunch. This animal is named Jessica, and she was rescued after floods in northern South Africa in 2000 left her orphaned. She was spotted stranded alone on the banks of the Blyde River, so park ranger Tonie Joubert elected to bring her home to his wife Shirley. History does not record Shirley's initial response, but currently Jessica seems to be a fully-fledged houseguest. Now this is the sort of thing that I normally run straight through the Hoax Slayers website. But the catch is, this is posted on the IUCN website's recent news reports section, after an article appeared in the *Daily Mail UK* online, 30 April 2008, so I guess it's one of those occasions when truth is stranger than fiction. But I just can't imagine the size of swimming pool and litter tray needed for such a companion animal.

RHINOCEROSES

There are five species of rhinoceros alive today, two in Africa and three in Asia. Far more information is available on the endangered status of these existing rhinos than there is on exactly how dangerous the individual animals are to individual human beings. They have been driven to the brink by poaching and habitat loss. It is true that the rhinoceros horn has been valued for dagger handles by Yemeni men, although other materials are now beginning to become acceptable. It is also true that powdered rhinoceros horn is highly valued in traditional Chinese medicine, mainly as a febrifuge (a drug to reduce fever) not as an aphrodisiac as is commonly asserted. Beginning with the cultural understanding that for some people this is considered a life-saving drug, some conservation programs are working to promote suitable substitutes as a way to prevent the total extinction of the rhinoceros. Given that during my research for this chapter I was constantly pestered by Internet advertisers hawking powdered rhinoceros horn on eBay, it can't happen soon enough. It is definitely a race against time.

The three Asian rhinoceroses are in the direst straits — the sum total population of all three species is less than that of the black rhinoceros of Africa, which is in none too good a shape itself. Of them all, the most critically endangered is the Sumatran rhino (*Dicerorhinus sumatrensis*), most of which have been lost to poaching, with only 250 to 400 individuals left in the wild. There are even fewer Javanese rhinos (*Dicerorhinus sondaicus*), but the

100 surviving individuals are considered to be slightly less imperilled as they live in a protected area of Indonesia. The Indian rhino (*Rhinoceros unicornis*) is in the best position of the three, having recovered to four-digit counts of 2000 plus. It's still an ominously low number.

Three of the four black rhinoceroses (*Diceros bicornis* spp.) of Africa are critically endangered. The fourth, the West African black rhinoceros, is probably extinct. The northern white rhinoceros (*Ceratotherium simum*) is set to join it. In fact, the only non-endangered rhinoceros is the southern subspecies of the white rhinoceros of Africa. (It is not white; the name derives from the Afrikaans word *wyd*, or wide — and refers to the size of its mouth.) And it, too, was once thought to be extinct. The rediscovery of several dozen white rhinos in 1895 marked the beginning of one of the most remarkable conservation success stories. Ironically, managed safari hunting on private game reserves has proved to be one of the key factors in rebuilding the numbers of this rhinoceros.

Given the rarity of these animals, then, most humans are likely to consider themselves vastly fortunate to see a wild rhinoceros at all, and are quite prepared to pay handsome sums for the opportunity to do so. That doesn't mean they want too close an encounter, however. It's still a very big and very dangerous animal. The heaviest white rhinoceros recorded was 4.5 tonnes and even the ordinary ones are heavy enough and measure between 1.5 and 1.8 metres tall. This is not an animal that you want running at you in a rage. Luckily, you have some chance of spotting a white rhino because they tend to be foraging more visibly than other rhino species in lower vegetation; consequently, they also have more chance of spotting you, which is a good thing all round. They are less aggressive than black rhinos and more likely to warn you off with a 'mock charge' or move away to a safer distance themselves.

THE SKIN THING

On the whole, rhinoceros skin is renowned for being so thick that it's like armour, hence the common expression 'he (or she) has the hide of a rhinoceros'. Only their mates and their physicians know that's not the whole story. A friend of mine who spent many years as a zoo veterinarian says that some parts of a rhino's body (the inside of the thigh being one) have skin as soft as any he's ever felt. One presumes the animal was anaesthetised when he made this discovery. The rhino is metaphorically thin-skinned, too. And they are built very strongly — the same friend has seen a zookeeper fly through the air with just a quick flick of a rhino's head. (The vet had some sympathy, having himself suffered a similar experience from a sly thwack of an elephant's trunk.)

Rhinos are like cats in that if they flatten their ears it's a clear signal that they're not happy! Unfortunately, in many cases people are so surprised when a black rhino charges out from behind cover that they have missed the opportunity to assess its body language. In 2003 an ornithologist surveying birds in the Royal Safari Chitwan National Park in Nepal was charged by a black rhino with a calf. Although injured, he survived. He was exceptionally lucky, as most such charges have fatal results, and there is at least one death a year from rhinoceroses in Chitwan alone, according to Bhimsen Thapaliya writing in *The Rising Nepal* in March 2003. Thapaliya explains that it happens most often in February when locals are allowed into the protected area to harvest *kans* grass for roof and fence material. Unfortunately, the tallness of the grass often prevents people seeing the rhinoceroses until it's too late.

BUFFALO

'Oh give me a home where the buffalo roam' was obviously not penned by an African farmer. Unlike its distant relative the Asian water buffalo, the African buffalo (*Syncerus caffer*) has not suffered itself to be domesticated, and is a very dangerous animal indeed. It has earned its ranking in 'the Big Five', alongside the African lion, African elephant, rhinoceros and leopard: the animals deemed most difficult and dangerous to hunt. It is still considered to be a premium game animal — some devotees even tackle them with bows. Hunting and habitat loss both threaten the species, but its relatively large population and wide distribution throughout central Africa means that, unlike other members of the Big Five, it is not yet an officially designated threatened species.

Whether or not the buffalo ranks as the most dangerous of all the Big Five is one of those perennial debates enjoyed by all parties. What is certain is that it can hold its own among other big predators. Only the largest and smartest of male lions stands a chance of bringing down a buffalo by itself, and it's a very risky business much better achieved by the whole pack. Leopards might stealthily pick off newborns, but that's as far as they'll venture. The Nile crocodile is capable of tackling the old and sick or young and vulnerable, but even they baulk at a full-grown healthy adult buffalo.

African buffalo seem to come in two basic sizes — large and enormous, although some taxonomists argue for a third subspecies, the Sudan buffalo which is halfway in size and colour between the two extremes. The smaller subspecies has a reddish coat and inhabits forested land. The black-coated savannah subspecies, often called the Cape buffalo, can be as tall as an average human (1.7 metres from the shoulder to the ground), but it weighs far more (averaging 700 kilograms, but up to 900 kilograms). Don't be misled by its bulk, however; it can almost give a greyhound a run for its money when it comes to swiftness. They have been clocked at speeds of just under 60 kilometres per hour. And all this power is fuelled by grass!

Exact numbers are elusive, but it is estimated that a dozen or more people are killed by African buffalo each year. Two widely reported deaths in recent years include that of Canadian hunter Bob Fontana, who was killed on 21 July 2004. An experienced hunter, 45-year-old Fontana was checking out a salt lick for game when the buffalo charged from behind some brush. Two years later an expatriate Englishman, Simon Combes, suffered a similar fate. Sixty-four-year-old Combes, a wildlife artist, was famous for his realistic close-up portraits of some of Africa's most dangerous wildlife. Reportedly, he was sightseeing in the Great Rift Valley with his wife and a friend when the animal charged — seemingly from nowhere — trampling him and then goring him in the chest.

These stories are characteristic: one of the main reasons the buffalo is judged to be extremely dangerous is because of the sheer unpredictability of its attack. No doubt many African people have died in like manner, but their deaths are seldom reported in the international media.

BISON

The buffalo song quoted earlier is, of course, a song of the American west, where the indigenous bison are commonly and wrongly referred to as buffalo. Originally one of six species, only the American (*Bison bison*) and the European (*B. bonasus*) bison remain. For many people bison are evocative of a vanished past; they were, after all, driven to near extinction in the United States. According to American environmental historian Andrew Isenberg, this had a lot to do with cultural domination of the Native American people by European settlers, as well as the usual motives of profit and protecting agricultural enterprise.

What few people know (although some locals have suspected) is that bison can be more of a problem for people than grizzly bears. A study in Yellowstone National Park found that over a decade or so from the late 1970s, the death toll from grizzlies and bison was the same: two each. Bison killed in 1971 and 1984. However, you were more than twice as likely to be attacked by a bison as you were a grizzly bear. This makes the bison by far the most dangerous animal in the park.

MOOSE

When I had the privilege of visiting Denali National Park in Alaska I was delighted to see free-ranging grizzly bears — from the safety of a bus! Outside the confines of a vehicle, I paid strict attention to the warnings about avoiding a closer encounter with a bear, and what to do if I did come face to face with one. So I was wandering along a walking trail by myself rehearsing this knowledge (don't run, look big) like some kind of mantra to avert danger, when the shadowy outline of a huge brown animal loomed out of the trees. Bloody hell! I was less than 3 metres away from a wild moose. We stood there eyeballing each other for about five minutes, then both of us wandered on our way. Prior to this experience, the closest I had ever been to a moose was watching the opening credits of *Northern Exposure*. How had I managed to miss the warnings about moose?

ELKISH ALLSORTS

Elk is a name that is applied to more than one species. In Europe the animal Americans call a moose (*Alces alces*) is sometimes referred to as a moose, but more commonly called an elk. Elk is also the name given to the large red deer of Britain, Europe and northern Africa (*Cervus elephus*). Its North American cousins, *Cervus canadensis* (also found in Asia) are often called wapiti. To top it off, the Indian sambar (*Cervus unicolor*) sometimes gets called an elk as well. As with most deer species, the primary danger presented by elks is via vehicle collisions.

There are two situations in which face-to-face encounters with moose (*Alces alces*) are particularly risky: a mother with calves will brook no perceived threat to her offspring and a rutting male will brook no perceived challenge to his macho supremacy. If you are charged and trampled by a moose, you will most certainly be seriously injured and possibly killed and the moose will walk away unharmed unless it is shot in retribution.

However, that is not the most likely way of dying by moose. The Alaskan moose (*Alces alces gigas*) is the largest deer in the world. It has earned the right to stand its ground. And so it does. It's an adaptation that serves it well in the day-to-day struggle for existence with the predator wolf. It is less useful when it comes to cars. Ironically, the coming of the highway to the northernmost regions of the United States was an open invitation to the moose, which moved into much closer contact with people. Some even moved into town: there are 450 resident moose in Anchorage and quite a few more come in for the winter. While this doesn't seem to have produced significant numbers of fatalities from direct encounters, it has produced a sizeable problem in terms of vehicle collisions, with occasional death tolls on both sides of the windscreen. Between 1991 and 1995 researchers reported a rise in MVCs (moose–vehicle collisions) in Anchorage. However, less than a fifth of these resulted in injuries (a total of 158 injured people). MVCs are much more likely at night, and the situation is more perilous on the open road. This is where most people die as a result of a high-speed encounter with a moose.

It's a situation that is familiar in northern Europe as well. In Finland, an estimated 20 per cent of all road accidents involve either moose or white-tailed deer, many resulting in serious injury, sometimes death. A smaller number of reindeer–vehicle collisions in northern Finland augur more ill for the reindeer than the people, as it is a smaller animal overall. (A comprehensive search turned up nil data about collisions with a big fat bearded guy in a red suit.)

ELEPHANTS

Recently there has been a reclassification of the African elephant into two separate species. *Loxodonta cyclotis* (an elusive and darker-skinned forest-dwelling group) was formerly considered a subspecies of *Loxodonta africana* (which lives on the savannah). However, DNA testing designed to trace poached ivory showed surprisingly substantial differences between the two. One thing remains inarguable: the savannah elephant is the world's largest living land mammal, with big males weighing in excess of 6 tonnes, and measuring close to 4 metres at shoulder height. (It's only beaten for the height record by its fellow African, the giraffe, the tallest of those on record having reached an extraordinary 6 metres.) In contrast, even the tallest Asian elephant (*Elephas maximus*) doesn't reach more than 3 metres (its highest point is the head, not the shoulder). Only exceptional Asian elephants attain the top weights of 5.5 tonnes, and most occupy the middle range down to a positively petite 2.25 tonnes. (The average modern LandCruiser is around 3.3 tonnes.)

Apart from the size discrepancy, there are a few other distinguishing features between African and Asian elephants. For a comparison of trunks, see 'Trunk call' on page 224. The ears of the Asian elephant and African forest elephant are rounded, whereas the ears of the African savannah elephant look like nothing so much as an outline of the African continent. The African elephant is the saggy-baggy one, and the Asian elephant is smooth-skinned. Both male and female African elephants are tuskers (the forest elephants' tusks are thinner and slightly pink), whereas the female Asian elephant only has 'tushes' (similar to tusks but much smaller and without nerves) which are seldom visible. The fabled 22-month pregnancy is the same for all species. Looked at from my puny human perspective it might take that long to gather up the courage to give birth to an infant that could weigh around 90 kilograms and be close to a metre tall!

TRUNK CALL

When it comes to their trunks, both species of elephant possess exactly the same amount of muscles, but one ends in a single finger (Asian) and the other in two (African). Trying to find out exactly how many muscles is a bit like finding out how many elephants can fit in a telephone box — a joke. Wild numbers are bandied about: 4000; 40,000; 60,000; 100,000 (do I hear any advance on 100,000?). The number is sixteen — eight each side. But don't expect any trivia or quiz night host to believe you. I suggest that on such occasions you have on your person the most informative scientific paper by Marchant and Shoshani, 'Head muscles of *Loxodonta africana* and *Elephas maximus* with comments on *Mammuthus primigenius* muscles'. Then you can prove beyond a shadow of a doubt that, while the Asian elephant's trunk might have 150,000 muscle fascicles (bundles) and the African elephant an unknown amount, they both possess the same modest number of actual muscles cited above.

It's quite evident that a creature so big poses a potential threat to humans, even inadvertently. Both Asian and African elephants can and do kill people, but the latter is considered to be more temperamentally likely to do harm. Comprehensive global statistics are simply not available, and the only way of gauging the extent of the problem is to look at specific regional tolls and extrapolate an average. So, for example, the total number of deaths recorded for North and South Bengal during the period 1990–97 was 451. In Bihar, south India, a combined total of 242 represents both fatalities and injuries caused by elephants between 1989 and 1994. In India overall, the best estimate is that between 100 and 200 people are killed by elephants each year. Not even this level of figures is available for the African subcontinent, although one Kenyan study revealed that elephants killed almost as many people as all the predator animals put together (221 elephant-caused fatalities compared with 250 for the combined total of predator-caused deaths).

A GRAVE PROBLEM

The whole world is familiar with the African myth of the elephant graveyard. From *King Solomon's Mines* to *The Lion King* we have been presented with the colourful story of a place known only to the elephants where the elderly pachyderms go to die, and bones and tusks pile up beyond measure. Such a mythically proportioned and elephant-chosen 'graveyard' has never been found.

Although the elephant's graveyard is mythical, it's not uncommon to find accumulations of bones from many dead elephants, most likely the last of a dying population that has descended on a shrinking food or water source and finally succumbed to starvation. Some hypothesise that this is the basis of the legend. And even though it is by no means proven that elephants visit the graveyard of their kin in the same way that humans do, that doesn't mean they don't honour and mourn their dead. It has been known anecdotally in Africa for a long time that elephants will pay a great deal of attention to the bones and ivory of dead elephants. In 2005, English scientist Karen McComb and her colleagues presented formal evidence of this. Elephants responded quite differently to elephant skulls and bones than to those of other species. This capacity to recognise and respond to the long-dead members of their race has only been found in one other species so far — humans. According to the researchers, '[The elephant's] interest in the ivory and skulls of their own species means that they would be highly likely to visit the bones of relatives who die within their home range.'

THE LONG MEMORY OF TRAUMA

For as long as elephants and people have occupied the same territory, there has been the danger of death from elephants, especially from males in must. On the whole, though, being killed by an elephant has only been one of the multitudes of far more common ways that people who live in elephant country can meet their end. That fact remains true, but something else, something more sinister in terms of human–elephant conflict seems to be happening today. Although exact figures are not forthcoming, a devastating trend is emerging. As the total numbers of elephants plummet, the number and severity of human–elephant conflicts rises correspondingly. At first it may seem odd that fewer numbers of animals are able to wreak so much extra damage. On reflection, one reason is obvious — the number of elephants is greatly reduced precisely because of habitat loss and fragmentation that brings them into far closer and more frequent contact with humans, with destructive consequences for both.

The second reason is not so obvious, but evidence is beginning to emerge of an even more chilling scenario. Humans have always at some level recognised the cross-generational human costs of traumas such as war and natural disaster. Still, it wasn't until 1980 that post-traumatic stress disorder (PTSD) was added to the US *Diagnostic and Statistical Manual of Mental Disorders*. Now some scientists are contending that the elephant is susceptible to this syndrome also. In a landmark essay published in *Nature* in 2005, researchers Bradshaw, Schore, Brown, Poole and Moss describe a killing rampage by teenage elephants that left more than 100 rhinoceroses dead. (False reports still circulate that cross-species rape was part of this rampage, due to the mistaken assumption that the mounting behaviour exhibited was about sexual penetration rather than a classic dominance–aggression strategy.)

The rampaging elephants were all orphans who had been rescued after poachers had killed their mothers and herd mates, and subsequently been moved to a sanctuary far away from their original home range. A decade later, the onset of puberty saw the eruption of this extraordinary violence. The researchers contend that this and other examples of hyper-aggression are accountable for by the long-term psychophysiological effects of trauma on both the brain and the personality. PTSD is a double-edged sword. Not only does it deeply affect the traumatised individual (both human and elephant) it also both reflects and perpetuates a serious breakdown of the social structure. The elephant is a highly intelligent animal that is very similar to human beings in terms of life span and social needs. The longevity of its memory is not a myth. The first MRI scan of an elephant brain was taken in 2008 (it can't have been in a standard machine). The scan showed that an elephant's brain possesses a huge hippocampus, which is the part of the mammalian brain that is a major storehouse of memory and processor of emotion.

It is deeply horrifying, but hardly surprising really, that elephants are aggressively manifesting the cumulative effects of profound ongoing trauma. Just as with humans, a great deal of this aggression is visited upon their own species. The *Nature* essay referred to a socially disrupted elephant community in a South African park where 90 per cent of all male elephants were killed by other male elephants. The average in unstressed elephant communities is 6 per cent. An article on elephants and PTSD by Charles Siebert was published in the *New York Times* in October 2006. Among the people he talked to was Ugandan animal ethologist Eve Abe, whose doctoral work drew striking comparisons between the fate of male elephants orphaned by war and poaching and the young male orphans of the Acholi, her own tribe. According to Abe, the unprecedented levels of violence that result from the unravelling of social structure are clearly mirrored in both human and elephant society, and she brings some convincing evidence to bear on her case.

A small fragment of hope can be found at the bottom of this Pandora's box of evils. The work done on diagnosing and treating PTSD in humans is now being applied to elephants with some signs of success. Of particular note is the work of the Sheldrick Trust, a Kenyan-based sanctuary for orphaned and traumatised elephants. By working within an 'elephant' world view, human 'allomothers' care for the young elephants, even sleep with them if necessary, for as long as is needed to restore some sense of security and trust. This mimics the structure of a natural herd where elephant 'allomothers' all contribute to the raising of the herd's infants. Nothing could be sweeter than the fact that when these traumatised orphan elephants have grown and been returned to wild herds they often bring back their wild-born offspring to visit the human 'allograndmother'. Perhaps if we can learn how to heal elephant societies, we stand a better chance of healing our own.

MUST RUMBLE

You must remember this: a bull elephant on a hormone high is in no mood for just kissing. Anyone sufficiently unlucky to be close enough to an unrestrained bull elephant in must may hear the legendary low bellowing known as the 'must rumble'. It may well be the last thing they ever hear. Because when a male elephant reaches the stage of must rumbling, they must, indeed, rumble. Their behaviour at this time resembles nothing so much as the hypothetical 'roid rage attributed to human body builders and weightlifters taking anabolic/androgenic steroids. And 'roid rage itself may be a symptom of increased testosterone, which is a singularly dangerous 'drug'.

The must elephant is so fuelled by testosterone — something in the order of 60 times the normal levels — that it not only vocalises menacingly, it also exudes secretions from the temporal gland between its eye and ear and constantly dribbles urine. Altogether, not a pleasant house-guest. Add in the fact that it is likely to kick your house down, either before or after trampling you and anyone or anything else that gets in its way, and you understand why the state of must is one of the primary causes of human–elephant, not to mention elephant–elephant, conflict.

Scientific understandings of must may help to reduce the dangers associated with this phenomenon. But an elephant in must is not exactly a cooperative research subject. Analysing the tarry stuff that comes from the temporal gland would present little challenge in the laboratory, for instance, but collecting samples of it certainly does. Apparently, even the production of this secretion is a painful business for the elephant — something akin to a root canal abscess. Not a time to get up close and personal in the interest of obtaining a sample.

Among the basic facts known about must are that the African

elephant is likely to start its cycles at around 25 years of age, the Asian elephant five years earlier. In both cases, often only the dominant male enters must, and less successful males either do not enter must at all, or if they do so, are less synchronised with the females' 14- to16-week oestrus cycle, hence are less successful breeders.

THE INDIAN ELEPHANT TEMPERANCE UNION

The spelling 'must' appears first in the *Oxford Dictionary*, followed by the common alternative 'musth'. The term comes from the Persian-derived Urdu word for intoxication, 'mast'. Ironically, S.S. Bist, West Bengal's Conservator of Forests, identifies illegal alcohol as a pertinent factor in some instances of human–elephant conflict. In a comprehensive report covering all aspects of the problem, Bist includes 'country liquor' as one of the issues to be addressed. Apparently, the forests adjacent to the tea gardens and villages of North Bengal are riddled with stills that attract wild elephants. And according to the no-nonsense Bist, 'a substantial number of people killed or injured by elephants are those who are intoxicated and cannot take care of themselves when confronted by an elephant'. Accordingly, one of his seven recommendations for promoting peaceable coexistence is as follows: 'People should not move outdoors after dusk in intoxicated condition and should also be warned against preparing and storing country liquor openly in their houses.'

Bist's other recommendations include the installation of electric lights in labourer's cottages, switching crops to less elephant-friendly food, and refraining from whitewashing cottages. Apparently, the elephants are less likely to attack a house that blends in to its environment by having earth-toned walls of ochre, green or dun. This is extremely useful advice for residents of North Bengal where more than 1000 houses are damaged by elephants each year.

MASS DESTRUCTION

The subject of poaching is one that arouses strong passions and fiercely held views, yet whichever way you look at it poaching is a tragedy of elephantine proportions. Unquestionably, along with habitat loss, the poaching of elephants for ivory has been one of the main drivers of the elephants' precipitous decline and fall in population numbers, from around a million wild elephants of both species 30 years ago, to less than half that number today. An estimated 20,000 elephants die each year to feed the illegal trade in ivory, much of which is now traded without regulation or sanction through the Internet.

And it's not only the elephants that are dying. Within a single month in the first half of 2007 seven wildlife rangers were killed by poachers in Tsavo (Kenya), Zakouma (Chad) and Virunga (Congo) national parks. It has to be one of the most high-risk occupations around. There is a tragic waste of human life on the other side of the equation, too, as the desperate struggle escalates and some regions have adopted a shoot to kill policy in what amounts to a 'war' for wildlife. Often those caught in this manner are amateurs and driven by desperation. It's sadly analogous to the trade in illegal drugs: shocking waste of life all around, and the poor and marginalised 'small fry' are the ones who pay the ultimate penalty while those making the obscene profits keep well away from the frontline and carry on business as usual.

And just when you think there are no further levels to this hell, you hear from Ugandan animal ethologist Eve Abe about one of the countless tragedies of the Ugandan–Tanzanian war which brought about the downfall of Idi Amin. By 1979, of 4000 elephants resident in Queen Elizabeth National Park, 150 remained. The rest had fallen to a poaching frenzy carried out by soldiers from both sides. In Abe's own words:

> *Normally when you say 'poaching', you think of people shooting one or two [elephants] and going off. But this was war. They'd just throw hand grenades at the elephants, bring whole families down and cut out the ivory. I call that mass destruction.*

AFTERWORD

MASS DESTRUCTION. THE PHRASE HAS BECOME SO COMMONPLACE, we hardly register the meaning anymore. Yet consider the accumulated weight of facts contained in *Deadly Beautiful*. Mass destruction is a precise description of this catalogue of worldwide loss of habitat, wholesale extinction of species. The deaths of individual humans have been accorded their dignified due in these pages. The far higher level of risk faced by people in developing countries has not been glossed over. Realistically, though, the combined impact of all the dangerous animals on humans as a species pales into insignificance when compared with the threat we pose to them.

At the end of the introduction, I spoke of two meanings of the word 'respect'. I will close the book with a third that shadows this subject. The phrase 'paying your respects' means the awkward platitudes we offer the bereaved in acknowledgement of those who have died. It is my heartfelt hope that *Deadly Beautiful* contributes to increasing practical knowledge of the actual level of threat posed by dangerous animals, as well as a profound respect for their intrinsic right to continue existing. Once bereft of these beautiful creatures, no amount of 'paying our respects' will bring them back.

Blessed are the extinct
for theirs was the kingdom of earth
Blessed are those who mourn
for there is no comfort to be sought.

— Liana Joy Christensen

Acknowledgements

I am deeply indebted to the following people who graciously consented to review chapter drafts, or contributed specialised information: Dr Phillip Arena, Dr Jeanette Covacevich, Dr Bill Gaynor, Adjunct Professor Barbara Main, OAM, Professor Brian Morton, Ms Becca Saunders, Professor Thomas S. Smith, Associate Professor Stephen J. Mullin, Professor Rick Shine and Ms Julianne Waldock. Their combined expertise has contributed greatly to the accuracy and inclusiveness of the book. Any remaining errors and omissions are strictly my responsibility.

Project manager Anouska Jones, editor Karen Gee and illustrator Ian Faulkner have all been fabulous to work with, deploying their various professional skills with brilliance and unfailing kindness.

My beloved husband Larry and my literary and non-literary friends have managed to live with (and often without) that most elusive and dangerous animal: an author with a deadline. Thanks to you all for your patience and understanding.

References

'Blood in the Surf', 2005, *People*, 11 July, 62–3. <http://0-www.proquest.com.prospero.murdoch.edu.au:80/> accessed 5 September 2007.

'Poisoning by venomous animals', 1967, *The American Journal of Medicine* (1), <http://www.sciencedirect.com/> accessed 13 May 2007.

International Consortium of Jellyfish Stings, ND, <http://medschool.umaryland.edu/ dermatology/jellyfish.asp> accessed 23 January 2008.

Hippopotamus Specialist Group of the IUCN, ND, <http://moray.ml.duke.edu/projects/hippos/> accessed circa June 2008

'Hospital sets deadly spider free', 2005, *ABC News Online*, <http://www.abc.net.au/news/newsitems/200505/s1357179.htm> accessed 9 September 2007.

Internet Center for Wildlife Damage Management. 2005, <http://icwdm.org/education/ProfessionalEducation.asp.> accessed 8 May 2008.

IUCN Red List of Threatened Species. ND, <http://www.iucnredlist.org/info/categories_criteria2001> accessed circa September 2007

FM 21 - 76 US Army Survival Manual, 2006, <http://www.humboldt.edu/~hsusnc/FM%2021-76%20US%20ARMY%20SURVIVAL%20MANUAL.pdf> accessed 8 July 2007.

Malaria control in complex emergencies: an inter-agency handbook, 2006, <http://www.who.int/malaria/docs/ ce_interagencyfhbook.pdf> accessed 3 May 2007.

Marine Medic, 2000, <http://www.marine-medic.com.au/> accessed circa December 2007

International Association for Bear Research and Management, 2007, <http://www.bearbiology.com/> accessed 7 April 2008.

New York Crime Rates 1960–2006, 2007, <http://www.disastercenter.com/crime/nycrime.htm> accessed 3 June 2008.

'Seven wildlife rangers killed', 2007, Briefly, *Oryx*, 41: 417–26, <doi:10.1017/S0030605307041427> accessed 8 June 2007

Shark attack File N D, <http://www.flmnh.ufl.edu/fish/Sharks/ISAF/ISAF.htm> accessed circa September 2007

'Snakes: Love 'em or leave 'em?', 2007, *Nature Watch*, <http://www.enature.com/articles/detail.asp?storyID=592> accessed 5 June 2007.

'WA: Shark attack victim punches animal', 2008, AAP General News Wire 11 May, <http://0-www.proquest.com. prospero.murdoch.edu.au:80/> accessed 4 June 2008.

'Wildlife Artist Simon Combes Killed by Cape Buffalo', 2005, <http://africanhuntinginfo.com/modules/news/index.php?storytopic=3&start=10> accessed 15 May 2008.

Adiguzel, S., Ozkan, O. and Inceoglu, B. 2007, 'Epidemiological and clinical characteristics of scorpionism in children in Sanliurfa, Turkey', *Toxicon* (6), <http://www.sciencedirect.com> accessed 22 May 2007.

Alam, M.I., Auddy, B. and Gomes, A. 1996, 'Viper venom neutralization by Indian medicinal plant (*Hemidesmus indicus* and *Pluchea indica* root extracts)', *Phytotherapy Research* (1), <http://dx.doi.org/10.1002/(SICI)1099-1573(199602)10:1<58::AID-PTR775>3.0.CO;2-F> accessed 19 June 2007.

Alkofahi, A., Sallal, A.J. and Disi, A.M. 1997, 'Effect of *Eryngium creticum* on the haemolytic activities of snake and scorpion venoms', *Phytotherapy Research* (7), <http://dx.doi.org/10.1002/(SICI)1099-1573(199711)11:7<540::AID-PTR150>3.0.CO;2-9> accessed 3 July 2007.

Azzoni, T. 2007, 'Brazilian saves grandson from anaconda' <http://www.washingtonpost.com/wp-dyn/content/article/2007/02/09/AR2007020900412.html> accessed 12 March 2008.

Bailey, P. and Wilce, J. 2001, 'Venom as a source of useful biologically active molecules', *Emergency Medicine* 13 (1):28–36, <http://www.ingentaconnect.com/content/bsc/emmold/2001/00000013/00000001/art00007 - aff_2#aff_2> accessed 3 July 2007.

Barklow, W.E. 2004, 'Amphibious communication with sound in hippos, *Hippopotamus amphibious*'. *Animal Behaviour* 68 (5): 1125–32, <http://www.sciencedirect.com/> accessed 5 June 2008.

Barnes, J.H. 1967, 'Extraction of cnidarian venom from living tentacle', *Toxicon* (4), <http://www.sciencedirect.com/> accessed 19 January 2008.

Bauer, H. and Van Der Merwe, S. 2004, 'Inventory of free-ranging lions', *Panthera leo* in Africa, *Oryx* 38 (1): 26–31, <http://journals.cambridge.org/> accessed 17 April 2008.

Beier, P. 1991, 'Cougar attacks on humans in the United States and Canada', *Wildlife Society Bulletin* 19: 403–12, <http://oak.ucc.nau.edu/pb1/vitae/Beier_1991.pdf> accessed 15 April 2008.

——1992, 'Cougar Attacks on Humans: An update and some further reflections', *Proceedings of the Fifteenth Vertebrate Pest Conference,* <http://digitalcommons.unl.edu/cgi/viewcontent.cgi?article=1005&context=vpc15> accessed 16 April 2007.

Bell, K.L., Sutherland, S.K. and Hodgson, W.C. 1998, 'Some pharmacological studies of venom from the inland taipan (*Oxyuranus microlepidotus*)', *Toxicon* (1), <http://www.sciencedirect.com/> accessed 31 May 2007.

Benchley, P. 2002, *Shark: True stories and lessons from the deep*, London, HarperCollins.

Bist, S.S. ND, 'Elephant–human conflict in West Bengal', <http://www.wii.gov.in/envis/bulletin/bist1.htm> accessed 19 May 2008.

Blaylock, R. 2004, 'Epidemiology of snakebite in Eshowe, KwaZulu-Natal, South Africa', *Toxicon* (2), <http://www.sciencedirect.com/> accessed 1 June 2007.

Bradshaw, G.A., Schore, A.N., Brown, J.L., Poole, J.H. and Moss, C.J. 2005, 'Elephant breakdown', *Nature* (7028), <http://il.proquest.com> accessed 11 June 2008.

Breitenmoser, U. 1998, 'Large predators in the Alps: The fall and rise of man's competitors', *Biological Conservation* (3), <http://www.sciencedirect.com/> accessed 8 June 2008.

Bruskotter, J.T., Schmidt, R.H. and Teel, T.L. 2007, 'Are attitudes toward wolves changing? A case study in Utah', *Biological Conservation* (1–2), <http://www.sciencedirect.com/> accessed 20 April 2008.

Burns, G.L. and Howard, P. 2003, 'When wildlife tourism goes wrong: a case study of stakeholder and management issues regarding dingoes on Fraser Island, Australia', *Tourism Management* (6), <http://www.sciencedirect.com/> accessed 5 September 2007.

Burridge, M.J, Simmons, L. and Condie, T. 2004, 'Control of an exotic tick (*Aponomma komodoense*) infestation in a Komodo dragon (*Varanus komodoensis*) exhibit at a zoo in Florida', *Journal of zoo and wildlife medicine: official publication of the American Association of Zoo Veterinarians* (2), <http://www.ncbi.nlm.nih.gov> accessed 2 July 2007.

Byard, R.W., Brown, K. and Gilbert, J. 1999, 'O10. Fatal shark attacks in South Australia — pathological features of two cases', *Journal of Clinical Forensic Medicine* 6 (3):188.

Calderon-Aranda, E.S., Dehesa-Davila, M., Chavez-Haro, A. and Possani, L. 1996, 'Scorpion stings in Mexico and their treatment', *Toxicon* (2), <http://www.sciencedirect.com/> accessed July 2, 2007.

Caldicott, D.G.E., Mahajani, R. and Kuhn, M. 2001, 'The anatomy of a shark attack: a case report and review of the literature', *Injury* (6), <http://www.sciencedirect.com/> accessed 1 November 2007.

Carrier, J.C, Musick, J.A. and Heithaus, M.R. (eds) 2004, *Biology of Sharks and their Relatives, CRC marine biology series*, Boca Raton, Fla., London, CRC.

Carwardine, M. and Watterson, K. 2002, *The Shark Watcher's Handbook: A guide to sharks and where to see them*, Princeton and Oxford, Princeton University Press.

Casey, S. 2005, *The Devil's Teeth*, Sydney, Pan Macmillan.

Chandy, S. and Euler, D.L. 2000 'Can community forestry conserve tigers in India?' *Personal, Societal, and Ecological Values of Wilderness: Sixth World Wilderness Congress proceedings on research, management and allocation, volume 11*: 155–61, <http://www.fs.fed.us/rm/pubs/rmrs_p014/rmrs_p014_155_161.pdf> accessed 18 April 2008.

Chippaux, J-P. 1998, 'The development and use of immunotherapy in Africa', *Toxicon* (11), <http://www.sciencedirect.com/> accessed 13 May 2007.

Chowell, G., Diaz-Duenas, P., Bustos-Saldana, R., Aleman Mireles, A. and Fet, V. 2006, 'Epidemiological and clinical characteristics of scorpionism in Colima, Mexico (2000–2001)', *Toxicon* (7), <http://www.sciencedirect.com> accessed 3 July 2007.

Christensen, L. J. 2006, 'Beastitudes' *Wild Familiars*, Fremantle , Tone River Press, p. 43.

Ciofi, C., Beaumont, M.A., Swingland, I.R. and Bruford, M.W. 1999, 'Genetic divergence and units for conservation in the Komodo dragon *Varanus komodoensis*', *Proceedings: Biological Sciences* (1435), <http://dx.doi.org/10.1098/rspb.1999.0918> accessed 14 August 2007.

— and Swingland, I.R. 1997, 'Environmental sex determination in reptiles', *Applied Animal Behaviour Science* (3–4), <http://www.sciencedirect.com/> accessed 30 August 2007.

Colombo-Tierramrica, F. 2007,'Environment: World's shark population down 80% in 15 years', *Global Information Network*, 12 March, <http://0-www.proquest.com.prospero. murdoch.edu.au:80/> accessed 4 July 2008.

Coombs, M.D., Dunachie, S.J., Brooker, S., Haynes, J., Church, J. and Warrell, D.A. 1997, 'Snake bites in Kenya: a preliminary survey of four areas', *Transactions of the Royal Society of Tropical Medicine and Hygiene* (3), <http://www.sciencedirect.com/> accessed 13 May 2007.

Corbett, L. 2001, *The Dingo in Australia and Asia*, Marleston, J.B. Books.

Corneille, M.G., Larson, S., Stewart, R.M., Dent, D., Myers, J.G., Lopez,

P.P., McFarland, M.J. and Cohn, S.M. 2006, 'A large single-center experience with treatment of patients with crotalid envenomations: outcomes with and evolution of antivenin therapy', *The American Journal of Surgery* (6), <http://www.sciencedirect.com/> accessed 31 May 2007.

Covacevich, J. 1994, *Dandarabilla* and *Gunjjiwuru*: The discovery of the taipans — the world's most dangerous snakes', *Milestones of Australian Medicine*, J. Pearn (ed.). Brisbane, The Australian Medical Association and Amphion Press, Dept. of Child Health Publishing Unit, University of Queensland.

——Davie, P. and Pearn, J. (eds) 1987, *Toxic Plants & Animals: A guide for Australia*, Brisbane, Queensland Museum.

Crachi, M.T., Hammer, L.W. and Hodgson, W.C. 1999, 'The effects of antivenom on the in vitro neurotoxicity of venoms from the taipans *Oxyuranus scutellatus, Oxyuranus microlepidotus and Oxyuranus scutellatus canni*', *Toxicon* (12), <http://www.sciencedirect.com/> accessed 5 June 2007.

Crome, F.H.J. and Moore, L.A. 1990, 'Cassowaries in North-Eastern Queensland — report of a survey and a review and assessment of their status and conservation and management needs', *Wildlife Research* (4), <http://www.publish.csiro.au/journals/> accessed 29 June 2008.

Cronin, K. 2006, '"The bears are plentiful and frequently good camera subjects", postcards and the framing of interspecies encounters in the Canadian Rockies', *Mosaic (Winnipeg)* (4), <http://infotrac.galegroup.com> accessed 1 December 2007.

Crowder, L.B. and Figueira, W.F. 2006, 'Chapter 15 — Marine Metapopulations', *Metapopulation Ecology and Marine Conservation*, J.P. Kritzer and P.F. Sale (eds), Burlington, Academic Press.

Cundall, D. and Greene, H.W. 2000, 'Feeding in snakes', *Feeding: Form, function and evolution in tetrapod vertebrates*, K. Schwenk (ed.), San Diego, Academic Press.

Currie, B.J. 2000, 'Clinical toxicology: a tropical Australian perspective', *Therapeutic Drug Monitoring* 22 (1):73–8.

——2006. Treatment of snakebite in Australia: the current evidence base and questions requiring collaborative multicentre prospective studies. *Toxicon* (7), http://www.sciencedirect.com/ (accessed May 29, 2007).

——and Jacups, S.P. 2005, 'Prospective study of *Chironex fleckeri* and other box jellyfish stings in the "Top End" of Australia's Northern Territory', *MJA* 183 (11/12): 631–36, http://www.mja.com.au/public/issues/183_11_051205/cur10057_fm.html> accessed 6 December 2007.

Davidson, J. 2005, 'Contesting stigma and contested emotions: personal experience and public perception of specific phobias', *Social Science & Medicine* (10), <http://www.sciencedirect.com/> accessed 4 May 2007.

Davidson, R. 1992, *Tracks*, Sydney, Vintage.

Davis, E.W. and Yost, Y.A. 1983, 'The ethnomedicine of the Waorani of Amazonian Ecuador', *Journal of Ethnopharmacology* (2), <http://www.sciencedirect.com/> accessed 1 July 2007.

De Giorgio, F., Rainio, J., Pascali, V.L. and Lalu, K. 2006, 'Bear attack — a unique fatality in Finland', *Forensic Science International*, <http://www.sciencedirect.com/> accessed 18 November 2007.

Dehesa-Davila, M. and Possani, L.D. 1994, 'Scorpionism and serotherapy in Mexico', *Toxicon* (9), <http://www.sciencedirect.com/> accessed 8 July 2007.

Diamond, J.M. 1987, 'Did Komodo dragons evolve to eat pygmy elephants?' *Nature* (6116), <http://dx.doi.org/10.1038/326832a0> accessed 8 August 2007.

Dimich-Ward, H., Rennie, D., Hartling, L., Guernsey, J.R., Pickett, W. and Brison, R.J. 2004, 'Gender differences in the occurrence of farm related injuries', *Occup Environ Med* 61 (1):52–6.

Dublin, H.T. and Hoare, R.E. 2004, 'Searching for solutions: the evolution of an integrated approach to understanding and mitigating human–elephant conflict in Africa', *Human Dimensions of Wildlife* (4), <http://www.informaworld.com/> accessed 3 June 2008.

Dudley, S.F.J. 1997, 'A comparison of the shark control programs of New South Wales and Queensland (Australia) and KwaZulu-Natal (South

Africa)', *Ocean & Coastal Management* (1), <http://www.sciencedirect.com/> accessed 1 October 2007.

Einterz, E.M. 2001, 'International aid and medical practice in the less-developed world: doing it right', *The Lancet* (9267), <http://www.sciencedirect.com/> accessed 28 May 2007.

El-Amin, E.O. 1992, 'Issues in management of scorpion sting in children', *Toxicon* (1), <http://www.sciencedirect.com/> accessed 4 July 2007.

Endom, E.E. 1995, 'Initial approach to the child who presents with bites or stings', *Seminars in Pediatric Infectious Diseases* (4), <http://www.sciencedirect.com/> accessed 9 June 2007.

Ernst, C.H. and Zug, G.R. 1996, *Snakes In Question: The Smithsonian answer book*, Melbourne, CSIRO Publishing.

Fenner, P.J. ND, 'The Global Problem of Cnidarian (Jellyfish) Stinging', M.D. thesis, University of London, < http://www.marine-medic.com.au/> accessed 2 November 2007.

Forster, L.M. 1992, 'The stereotyped behaviour of sexual cannibalism in *Latrodectus hasselti* Thorell (Araneae: Theridiidae), the Australian redback spider', *Australian Journal of Zoology* (1), <http://www.publish.csiro.au/ journals/> accessed 17 July 2007.

Fowler, M.E. 1974, 'Diseases of children acquired from nondomestic animals', *Current Problems in Pediatrics* (10), <http://www.sciencedirect.com/> accessed 30 November 2007.

Freire-Maia, L., Campos, J.A. and Amaral, C.F.S. 1994, 'Approaches to the treatment of scorpion envenoming, *Toxicon* (9), <http://www.sciencedirect.com/> accessed 2 July 2007.

——1996, 'Treatment of scorpion envenomation in Brazil', *Toxicon* (2), <http://www.sciencedirect.com/> accessed 3 July 2007.

Garnet, J.R. (ed.) 1972, *Venomous Australian Animals Dangerous to Man*, 3rd edn, Parkville, Commonwealth Serum Laboratories.

Garrett, L.C. and Conway, G.A. 1999, 'Characteristics of moose–vehicle collisions in Anchorage, Alaska, 1991–1995', *Journal of Safety Research* (4), <http://www.sciencedirect.com/> accessed 21 March 2008.

Gazarian, K.G., Gazarian, T., Hernandez, R. and Possani, L.D. 2005,

'Immunology of scorpion toxins and perspectives for generation of anti-venom vaccines', *Vaccine* (26), <http://www.sciencedirect.com/> accessed 31 May 2007.

Gold, B.S. and Pyle, P. 1998, 'Successful treatment of neurotoxic King Cobra envenomation in Myrtle Beach, South Carolina', *Annals of Emergency Medicine* (6), <http://www.sciencedirect.com/> accessed 29 May 2007.

Gopalakrishnakone, P. 1990, 'A computer based colour-photo database system for dangerous animals and plants: academic and public information networks', *Toxicon*, 28:11 1285–92.

Graham, A. 1990, *Eyelids of Morning: The mingled destinies of crocodiles and men*, San Francisco, Chronicle Books.

Greene, H.W. 1997, *Snakes the Evolution of Mystery in Nature*, Berkeley, University of California Press.

Grundy, J.H. 1979, *Medical Zoology for Travellers*, 3rd edn, Chilbolton, Noble.

Gutiérrez, J.M., Theakston, R.D.G. and Warrell, D.A. 2006, 'Confronting the neglected problem of snake bite envenoming: the need for a global partnership', *PLoS Medicine* (6), <http://dx.doi.org/10.1371%2Fjournal.pmed.0030150> accessed 1 June 2007.

Hartman, L.J. and Sutherland, S.K. 1984, 'Funnel-web spider *Atrax robustus* antivenom in the treatment of human envenomation', *MJA* 141 (12–13):796–9.

Hawgood, B.J. 2006, 'The marine biologist — Bob Endean', *Toxicon* (7), <http://www.sciencedirect.com/> accessed 3 December 2007.

Hecht, M.K. and Marien, D. 1956, 'The coral snake mimic problem: a reinterpretation', *Journal of Morphology* (2), <http://dx.doi.org/10.1002/jmor.1050980207> accessed 29 May 2007.

Hedlund, J.H., Curtis, P.D., Curtis, G. and Williams, A.F. 2004, 'Methods to reduce traffic crashes involving deer: What works and what does not', *Traffic Injury Prevention* (2), <http://www.informaworld.com/> accessed 3 June 2008.

Heikkila, M. 2005, 'Alaska overrules voters, reinstates aerial wolf hunt',

<http://www.guerrillanews.com/articles/1430/Alaska_Overrules_Voters_Reinstates_Aerial_Wolf_Hunt> accessed 3 May 2008.

Heneman, B. and Glazer, M. 1996, 'More rare than dangerous: a case study of white shark conservation in California', *Great White Sharks*, A.P. Klimley and D.G. Ainley (eds), San Diego, Academic Press.

Herrero, S. and Fleck, S. 1990, 'Injury to people inflicted by black, grizzly or polar bears: recent trends and new insights', *Bears. Their Biology and Management. Papers from a conference, Alberta, B.C.*, S. Herrero (ed.), Morges, Switzerland, International Union for Conservation of Nature and Natural Resources.

Hoffman, D.R. 1998, 'Venoms', *Encyclopedia of Immunology*, P.J. Delves (ed.). Oxford, Elsevier.

Howard, R.J. and Burgess, G.H. 1993, 'Surgical hazards posed by marine and freshwater animals in Florida', *The American Journal of Surgery* (5), <http://www.sciencedirect.com/> accessed 14 November 2007.

Huffman, B. 2008, *Syncerus caffer* African buffalo, <http://www.ultimateungulate.com/Artiodactyla/Syncerus_caffer.html> accessed 2 June 2008.

Huggler, J. 2006, 'Animal behaviour: Rogue elephants', *The Independent — Environment*, <http://www.independent.co.uk/ environment/animal-behaviour-rogue-elephants-419678.html> accessed 19 May 2008.

Hung, D-Z. 2004, 'Taiwan's venomous snakebite: epidemiological, evolution and geographic differences', *Transactions of the Royal Society of Tropical Medicine and Hygiene* (2), <http://www.sciencedirect.com/> accessed 2 June 2007.

Hunt, M., Bylsma, L., Brock, J., Fenton, M., Goldberg, A., Miller, R., Tran, T. and Urgelles, J. 2006, 'The role of imagery in the maintenance and treatment of snake fear', *Journal of Behavior Therapy and Experimental Psychiatry* (4), <http://www.sciencedirect.com/> accessed 3 May 2007.

Isbell, L. A. 2006, 'Snakes as agents of evolutionary change in primate brains', *Journal of Human Evolution* (1), <http://www.sciencedirect.com/> accessed 8 May 2007.

Isbister, G.K. 2001, 'Venomous fish stings in tropical Northern Australia', *American Journal of Emergency Medicine* (7), <http://www.sciencedirect.com/> accessed 29 October 2007.

Iscan, M.Y. and McCabe, B.Q. 2000, 'ANTHROPOLOGY| Animal Effects on Human Remains', *Encyclopedia of Forensic Sciences*, J.A. Siegel (ed.). Oxford, Elsevier.

Ismail, M. 1995, 'The scorpion envenoming syndrome', *Toxicon* (7), <http://www.sciencedirect.com/> accessed 2 July 2007.

——and Memish, Z.A. 2003, 'Venomous snakes of Saudi Arabia and the Middle East: a keynote for travellers', *International Journal of Antimicrobial Agents* (2), <http://www.sciencedirect.com/> accessed 21 May 2007.

Jessop, T.S., Madsen, T., Ciofi, C., Jeri Imansyah, M., Purwandana, D., Rudiharto, H., Arifiandy, A. and Phillips, J.A. 2007, 'Island differences in population size structure and catch per unit effort and their conservation implications for Komodo dragons', *Biological Conservation* (2), <http://www.sciencedirect.com/> accessed 31 July 2007.

Jimenez-Ferrer, J.E., Perez-Teran, Y.Y., Roman-Ramos, R. and Tortoriello, J. 2005, 'Antitoxin activity of plants used in Mexican traditional medicine against scorpion poisoning', *Phytomedicine* (1-2), <http://www.sciencedirect.com/> accessed 10 July 2007.

Jones, P. 1991, 'South African shark conservation', *Marine Pollution Bulletin* (12), <http://www.sciencedirect.com/> accessed 22 October 2007.

Jorge da Silva, N. and Aird, S.D. 2001, 'Prey specificity, comparative lethality and compositional differences of coral snake venoms', *Comparative Biochemistry and Physiology Part C: Toxicology & Pharmacology* (3), <http://www.sciencedirect.com/> accessed 3 May 2007.

Kaltenborn, B.P., Bjerke, T., Nyahongo, J.W. and Williams, D.R. 2006, 'Animal preferences and acceptability of wildlife management actions around Serengeti National Park, Tanzania', *Biodiversity and Conservation* 15 (14):4633–49.

Kareiva, P. 2002, 'In brief', *Trends in Ecology and Evolution* (9),

<http://www.sciencedirect.com/> accessed 2 August 2007.

Karlsson, J. and Sjöström, M. 2007, 'Human attitudes towards wolves, a matter of distance', *Biological Conservation* (4), <http://www.sciencedirect.com/> accessed 17 May 2008.

Kay, W.R. 2005, 'Population ecology of *Crocodylus porosus* (Schnedider 1801) in the Kimberley region of Western Australia' PhD, School of Integrative Biology, University of Queensland, Brisbane, <http://espace.library.uq.edu.au/ view/UQ:9640> accessed 26 October 2007.

Khattak, A.J. 2003, 'Human fatalities in animal-related highway crashes', *Transportation Research Record* (1840):158–66.

Krisp, J.M. and Durot, S. 2007, 'Segmentation of lines based on point densities — an optimisation of wildlife warning sign placement in southern Finland', *Accident Analysis & Prevention* (1), <http://www.sciencedirect.com/> accessed 1 June 2008.

Laing, G.D., Renjifo, J.M., Ruiz, F., Harrison, R.A., Nasidi, A., Gutiérrez, J.M., Rowley, P.D., Warrell, D.A. and Theakston, R.D.G. 2003, 'A new pan African polyspecific antivenom developed in response to the antivenom crisis in Africa', *Toxicon* (1), <http://www.sciencedirect.com/science/> accessed 7 June 2007.

Lalloo, D. 2005, 'Venomous bites and stings', *Medicine (Tropical infections 2)* (8), <http://www.sciencedirect.com/> accessed 1 August 2007.

Lalloo, D.G., Trevett, A.J., Saweri, A., Naraqi, S., Theakston, R.D.G. and Warrell, D.A. 1995, 'The epidemiology of snake bite in Central Province and National Capital District, Papua New Guinea', *Transactions of the Royal Society of Tropical Medicine and Hygiene* (2), <http://www.sciencedirect.com/> accessed 21 May 2007.

Langley, R.L. 2005, 'Animal-related fatalities in the United States — an update', *Wilderness & Environmental Medicine [NLM — MEDLINE]* (2), <http://il.proquest.com> accessed 10 May 2007.

——and Morrow, W.E. 1997, 'Deaths resulting from animal attacks in the United States', *Wilderness and Environmental Medicine* 8 (1):8–16.

——and Hunter, J.L. 2001, 'Occupational fatalities due to animal-related events', *Wilderness and Environmental Medicine* 12: 168–74.

Leggat, P.A., Harrison, S.L., Fenner, P.J., Durrheim, D.N. and Swinbourne, A.L. 2005, 'Health advice obtained by tourists travelling to Magnetic Island: a risk area for "Irukandji" jellyfish in North Queensland, Australia', *Travel Medicine and Infectious Disease*, 3 (2):27–31, <http://www.sciencedirect.com/> accessed 29 October 2005.

Lehane, L. and Lewis, R.J. 2000, 'Ciguatera: recent advances but the risk remains', *International Journal of Food Microbiology* (2–3), <http://www.sciencedirect.com/> accessed 2 December 2007.

Lopez, B.H. 1977, *Giving Birth to Thunder: Sleeping with his Daughter*, New York, Avon Books.

——1978, *Of Wolves and Men*, New York, MacMillan Publishing Company.

Maine Professional Guides Association. 2002, <http://www.maineguides.org/referendum/pdf/Bear_Attacks.pdf> accessed 5 November 2007.

Marc, P., Canard, A. and Ysnel, F. 1999, 'Spiders (Araneae) useful for pest limitation and bioindication', Agriculture, *Ecosystems & Environment* (1–3), <http://www.sciencedirect.com/> accessed 13 June 2007.

Marchant, G.H. and Shoshani, J. 2007, 'Head muscles of *Loxodonta africana* and *Elephas maximus* with comments on *Mammuthus primigenius* muscles', *Quaternary International (World of Elephants 2 — Selected papers from the 2nd Congress, Mammoth Site of Hot Springs, 2nd World of Elephants Congress)*, <http://www.sciencedirect.com/> accessed 1 June 2008.

Maretic, Z. 1982, 'Some clinical and epidemiological problems of venom poisoning today', *Toxicon* 20 (1):345–8.

Martin, R.A. 2007, 'A review of shark agonistic displays: comparison of display features and implications for shark–human interactions', *Marine and Freshwater Behaviour and Physiology* (1), <http://www.informaworld.com/> accessed 14 September 2007.

Martz, W. 1992, 'Plants with a reputation against snakebite', *Toxicon* (10), <http://www.sciencedirect.com/> accessed 2 June 2007.

McGregor, J. 2005, 'Crocodile crimes: people versus wildlife and the politics of postcolonial conservation on Lake Kariba, Zimbabwe', *Geoforum* (3), <http://www.sciencedirect.com/> accessed 29 August 2007.

Merchant, M.E., Pallansch, M., Paulman, R.L., Wells, J.B., Nalca, A. and Ptak, R. 2005, 'Antiviral activity of serum from the American alligator (*Alligator mississippiensis*)', *Antiviral Research* (1), <http://www.sciencedirect.com/> accessed 29 August 2007.

Mirtschin, P. 2006, 'The pioneers of venom production for Australian antivenoms', *Toxicon* (7), <http://www.sciencedirect.com/> accessed 2 June 2007.

Mitchell, P.B. 1987, 'Here be Komodo dragons', *Nature* (6135), <http://dx.doi.org/10.1038/329111a0> accessed 2 September 2007.

Mode, N.A., Hackett, E.J. and Conway, G.A. 2005, 'Unique occupational hazards of Alaska: animal-related injuries', *Wilderness & Environmental Medicine* (4), <http://il.proquest.com> accessed 21 June 2008.

Molesworth, A.M., Harrison, R., Theakston, R.D.G. and Lalloo, D.G. 2003, 'Geographic information system mapping of snakebite incidence in northern Ghana and Nigeria using environmental indicators: a preliminary study', *Transactions of the Royal Society of Tropical Medicine and Hygiene (Society Meeting)* (2), <http://www.sciencedirect.com/> accessed 3 May 2007.

Montgomery, J.M., Gillespie, D., Sastrawan, P., Fredeking, T.M. and Stewart, G.L. 2002, 'Aerobic salivary bacteria in wild and captive Komodo dragons', *Journal of wildlife diseases (3),* <http://www.ncbi.nlm.nih.gov/entrez/query.fcgi?cmd=Retrieve&db=pubmed&dopt=Abstract&list_uids=12238371> accessed 22 August 2007.

Morton, B. 2005, 'Crocodiles and sharks', *Marine Pollution Bulletin* (5), <http://www.sciencedirect.com/> accessed 2 June 2007.

Mullen, G.R. and Stockwell, S.A. 2002, 'Scorpions (Scorpiones)', *Medical and Veterinary Entomology*, G. Mullen and L. Durden (eds), San Diego, Academic Press.

Murphy, R.C. 1996, 'Chapter 2 — A plea for white shark conservation', *Great White Sharks*, A.P. Klimley and D.G. Ainley (eds), San Diego,

Academic Press.

Musick, J.A., Carrier, J.C.and Heithaus, M.R. 2004, *Biology of sharks and their relatives, CRC marine biology series*, Boca Raton, Fla., London, CRC.

Nahum, A. and Kochva, E. 1974, 'The quantities of venom injected into prey of different size by *Vipera palaestinae* in a single bite', *Journal of Experimental Zoology* (1), <http://dx.doi.org/10.1002/jez.1401880108> accessed 25 May 2007.

Nelson, L. 2004, 'Venomous snails: one slip and you're dead', *Nature* 429: 798–9, <http://www.nature.com/nature/journal/v429/n6994/full/429798a.html> accessed 23 October 2007.

Nhachi, C.F.B. and Kasilo, O.M.J. 1993, 'Poisoning due to insect and scorpion stings/bites', *Hum Exp Toxicol.* 12 (2):123–5.

Nicholson, G.M., Graudins, A., Wilson, H.I., Little, M. and Broady, K.W. 2006, 'Arachnid toxinology in Australia: from clinical toxicology to potential applications', *Toxicon* (7), <http://www.sciencedirect.com/> accessed 25 July 2007.

Njau, J.K. and Blumenschine, R.J. 2006, 'A diagnosis of crocodile feeding traces on larger mammal bone, with fossil examples from the Plio-Pleistocene Olduvai Basin, Tanzania', *Journal of Human Evolution* 50 (2): 142–62, <http://www.sciencedirect.com/> accessed 20 November, 2007.

Nowell, K. 2001, 'Status and Conservation of the Felidae', *Handbook of the Mammals of the World, Volume 1: Carnivores*, Lynx Edicions, in press, <http://www.felidae.org/LIBRARY/Lynx%20Ediciones%20Felidae%20Status%20and%20Conservation%20Chapter.doc> accessed 20 April 2008.

Nowak, R. 2007, 'This shark is telling you something', *New Scientist*, 28 April, 12–13, <http://0-www.proquest.com.prospero.murdoch.edu.au:80/> accessed 4 September 2007.

Nyhus, P.J., Tilson, R.L. and Tomlinson, J.L. 2003, 'Dangerous animals in captivity: *ex situ* tiger conflict and implications for private ownership of exotic animals', *Zoo Biology* 22:573–86,

<http://dx.doi.org/10.1002/zoo.10117> accessed 3 June 2008.

Oakman, B. 2001, 'The problems with keeping dingoes as pets and dingo conservation', *A Symposium on the Dingo*, 34–38. C.R. Dickman and D. Lunney (eds), Sydney: Royal Zoological Society of NSW.

Okello, M.M., Manka, S.G. and D'Amour, D.E. 2008, 'The relative importance of large mammal species for tourism in Amboseli National Park, Kenya', *Tourism Management* (4), <http://www.sciencedirect.com/> accessed 8 June 2008.

Orams, M.B. 2002, 'Feeding wildlife as a tourism attraction: a review of issues and impacts', *Tourism Management* (3), <http://www.sciencedirect.com/> accessed 6 June 2008.

O'Shea, M. 2005. *Venomous Snakes of the World*, London, New Holland.

Otway, N.M., Bradshaw, C.J.A. and Harcourt, R.G. 2004, 'Estimating the rate of quasi-extinction of the Australian grey nurse shark (*Carcharias taurus*) population using deterministic age- and stage-classified models', *Biological Conservation* (3), <http://www.sciencedirect.com/science/> accessed 29 October 2005.

Ozanne-Smith, J., Ashby, K. and Stathakis, V.Z. 2001, 'Dog bite and injury prevention — analysis, critical review, and research agenda', *Injury Prevention* (4) <http://injuryprevention.bmj.com/cgi/content/abstract/7/4/321> accessed 29 May 2008.

Pearce, A. 2002, 'The cull of the wild — dingoes, development and death in an Australian tourist location', *Anthropology Today* 18 (5): 14–19, <http://www3.interscience.wiley.com/> accessed 15 April 2008.

Pearn, J. and Covacevich, J. (eds) 1988, *Venoms & Victims*, South Brisbane, Queensland Museum and Amphion Press.

Pearn, J., Covacevich, J. and Winkel, K.D. 2006, 'Toxinology in Australia's colonial era: a chronology and perspective of human envenomation in 19th century Australia', *Toxicon* (7), <http://www.sciencedirect.com/> accessed 12 May 2007.

Pennington, C. 2006, 'Tick tactics — Stephen Wikel takes aim at a tiny but powerful foe', <http://grad.uchc.edu/current/ pdfs/wikelsumm06.pdf.>

accessed 13 July 2007.

Peterson, M.E. 2006, 'Snake bite: pit vipers', *Clinical Techniques in Small Animal Practice (Practical Toxicology)* (4), <http://www.sciencedirect.com/> accessed 22 May 2007.

Platnick, N.I. 2008, The world spider catalog, version 8.5.: American Museum of Natural History.

Plumwood, V. 1999, 'Being prey', *The New Earth Reader: The best of Terra Nova*, D. Rothenberg and M. Ulvaeus (eds). Cambridge, Mass., MIT Press.

Pugh, R.N.H., Bourdillon, C.C.M., Theakston, R.D.G. and Reid, H.A. 1979, 'Bites by the carpet viper in the Niger Valley', *The Lancet* (8143), <http://www.sciencedirect.com/> accessed 17 May 2007.

Pugh, R.N.H., Bourdillon, C.C.M. and Theakston, R.D.G. 1980, 'Incidence and mortality of snake bite in Savanna Nigeria', *The Lancet* (8205), <http://www.sciencedirect.com/> accessed 22 May 2007.

Quammen, D. 2003, Shadow of the Nine-toed Bear', *Monster of God*, New York, W.W. Norton and Company.

Quigley, H. and Herrero, S. 2005, 'Characterization and prevention of attacks on humans', *People and Wildlife: Conflict or co-existence?*, R. Woodroffe, S. Thirgood and A. Rabinowitz (eds), Cambridge, Cambridge University Press.

Reid, H.A. 1956, 'Paper: sea-snake bite research', *Transactions of the Royal Society of Tropical Medicine and Hygiene* (6), <http://www.sciencedirect.com/> accessed 19 May 2007.

Ripple, W.J., and Beschta, R.L. 2007, 'Restoring Yellowstone's aspen with wolves', *Biological Conservation* (3–4), <http://www.sciencedirect.com/> accessed 2 June 2008.

Roskaft, E., Bjerke, T., Kaltenborn, B., Linnell, J.D.C and Andersen, R. 2003, 'Patterns of self-reported fear towards large carnivores among the Norwegian public', *Evolution and Human Behavior* (3), <http://www.sciencedirect.com/> accessed 2 June 2008.

Ross, M.C. 2001, *Dangerous Beauty: Life and Death in Africa: True stories from a safari guide*. New York, Hyperion.

Russell, F.E. 2001, 'Chapter 26 — Toxic effects of terrestrial animal venoms and poisons', *Toxicology: The basic science of poisons*, L.J. Casarett, C.D. Klaassen and J. Doull (eds), McGraw-Hill Professional.

——Walter, F.G., Bey, T.A. and Fernandez, M.C. 1997, 'Snakes and snakebite in Central America', *Toxicon* (10), <http://www.sciencedirect.com/> accessed 30 May 2007.

Sahi, A. 2008, 'The tale of the tiger never told', *Tehelka Magazine*, 5 (14). <http://www.tehelka.com/story_main38.asp?filename=Ne120408tale_tiger.asp> accessed 15 April 2008.

Sasa, M. and Vazquez, S. 2003, 'Snakebite envenomation in Costa Rica: a revision of incidence in the decade 1990–2000', *Toxicon* (1), <http://www.sciencedirect.com/> accessed 21 May 2007.

Savolainen, P., Leitner, T., Wilton, A.N., Matisoo-Smith, E. and Lundeberg, J. 2004, 'A detailed picture of the origin of the Australian dingo, obtained from the study of mitochondrial DNA', *Proceedings of the National Academy of Sciences* (33), <http://www.pnas.org/cgi/content/abstract/101/33/12387> accessed 23 May 2008.

Siebert, C. 2006, 'An elephant crackup?' *New York Times Magazine*, <http://www.nytimes.com/2006/10/08/ magazine/08elephant.html> accessed 19 May 2008.

Smith, T.S., and Herrero, S. 2007, 'A century of bear–human conflict in Alaska: analyses and implications', <http://www.absc.usgs.gov/research/brownbears/attacks/bear-human_conflicts.htm> accessed 14 December 2007.

Sollod, B.L., Wilson, D., Zhaxybayeva, O., Gogarten, J.P., Drinkwater R. and King G. F. 2005, Were arachnids the first to use combinatorial peptide libraries? *Peptides (Invertebrate Neuropeptides V)* (1), <http://www.sciencedirect.com/> (accessed July 2, 2007).

Spira, A.M. 2006, 'Preventive guidance for travel: trauma avoidance and medical evacuation', *Disease-a-Month (Travel Medicine, Part I)* (7), <http://www.sciencedirect.com/> accessed 3 September 2007.

Steiner, R. 2001, 'The tiger's best friend', *International Wildlife*, 31 (6),

12–25, <http://il.proquest.com> accessed 15 April 2008.

Stevens, J.D. 2007, 'Whale shark (*Rhincodon typus*) biology and ecology: a review of the primary literature', *Fisheries Research (Whale Sharks: Science, Conservation and Management — Proceedings of the First International Whale Shark Conference)* (1), <http://www.sciencedirect.com/> accessed 21 November 2007.

Stewart, C.J. 2003, 'Snake bite in Australia: first aid and envenomation management', *Accident and Emergency Nursing* (2), <http://www.sciencedirect.com/> accessed 4 May 2007.

Stock, R.P., Massougbodji, A., Alejandro, A. and Chippaux, J-P. 2007, 'Bringing antivenoms to Sub-Saharan Africa', *Nature Biotechnology* (2), <http://il.proquest.com> accessed 15 November 2007.

Straight, R.C. and Glenn, J.L. 'Human fatalities caused by venomous animals in Utah, 1900–90', *Great Basin Nat.*, 53 (4), 390–3.

Tedford, H.W., Sollod, B.L., Maggio, F. and King, G.F. 2004, 'Australian funnel-web spiders: master insecticide chemists', *Toxicon* (5), <http://www.sciencedirect.com/> accessed 3 July 2007.

Thakur, J.S., Mohan, C. and Sharma, D.R. 2007, 'Himalayan black bear mauling: offense or defense?' *American Journal of Otolaryngology*, 16 April, <http://www.sciencedirect.com/> accessed 9 November 2007.

Theakston, R.D.G. 1989, 'Snake venoms in science and clinical medicine 2. Applied immunology in snake venom research', *Transactions of the Royal Society of Tropical Medicine and Hygiene* (6), <http://www.sciencedirect.com/> accessed 11 May 2007.

——1995, 'Screening of antivenoms', *Toxicon* (5), <http://www.sciencedirect.com/> accessed 19 May 2007.

——1996, 'The kinetics of snakebite envenoming and therapy', *Toxicon* (2), <http://www.sciencedirect.com/> accessed 15 May 2007.

——and Warrell, D.A. 1995, '1st international congress on envenomations and their treatments: Institut Pasteur, Paris, 7–9 June, 1995', *Biologicals* (4), <http://www.sciencedirect.com/> accessed 2 May 2007.

——Warrell, D.A. and Griffiths, E. 2003, 'Report of a WHO workshop on the standardization and control of antivenoms', *Toxicon* (5),

<http://www.sciencedirect.com/> accessed 2 June 2007.

Thorbjarnarson, J., Mazzotti, F., Sanderson, E., Buitrago, F., Lazcano, M., Minkowski, K., Muniz, M., Ponce, P., Sigler, L., Soberon, R., Trelancia, A.M. and Velasco, A. 2006, 'Regional habitat conservation priorities for the American crocodile', *Biological Conservation* (1), <http://www.sciencedirect.com/> accessed 29 August 2007.

Thorbjarnarson, J., Wang, X., Ming, S., He, L., Ding, Y., Wu, Y. and McMurry S.T. 2002, 'Wild populations of the Chinese alligator approach extinction', *Biological Conservation* (1), <http://www.sciencedirect.com/> accessed 29 August 2007.

Tian, J., Paquette-Straub, C., Sage, E.H., Funk, S.E., Patel, V., Galileo, D. and McLane, M.A. 2007, 'Inhibition of melanoma cell motility by the snake venom disintegrin eristostatin', *Toxicon* (7), <http://www.sciencedirect.com/> accessed 29 May 2007.

Tibballs, J. 2006, 'Australian venomous jellyfish, envenomation syndromes, toxins and therapy', *Toxicon* (7), <http://www.sciencedirect.com/> accessed 29 January 2008.

——2006, 'Struan Sutherland — Doyen of envenomation in Australia', *Toxicon* (7), <http://www.sciencedirect.com/> accessed 2 June 2007.

Timm, R.M., Baker, R.O., Bennett, J.R. and Coolahan, C.C. 2004, 'Coyote attacks: an increasing suburban problem*', Proceedings of the 21st Vertebrate Pest Conference, 47–57, <http://repositories.cdlib.org/cgi/viewcontent.cgi?article=1004&context=anrrec/hrec> accessed 1 May 2008.

Toovey, S. 2006, 'Travelling to Africa: health risks reviewed', *Travel Medicine and Infectious Disease (Reviews: Essentials of Travel Medicine)* (3–4), <http://www.sciencedirect.com/> accessed 30 January 2008.

Trape, J.F., Pison, G., Guyavarch, E. and Mane, Y. 2001, 'High mortality from snakebite in south-eastern Senegal', *Transactions of the Royal Society of Tropical Medicine and Hygiene* (4), <http://www.sciencedirect.com/> accessed 16 May 2007.

Treves, A. and Naughton-Treves, L. 1999, 'Risk and opportunity for humans coexisting with large carnivores', *Journal of Human Evolution* 36

(3): 275–82, <http://www.sciencedirect.com/> accessed 4 April 2008.

Tricas, T.C., Deacon, K., Last, P., McCosker, J.E., Walker, T.I. and Taylor, L. 1997, *Collins Sharks and Rays: The ultimate guide to underwater predators*, L.R. Taylor (ed.), The Nature Company Guides, London, HarperCollins.

Ushanandini, S., Nagaraju, S., Kumar, K.H., Vedavathi, M., Machiah, D.K., Kemparaju, K., Vishwanath, B.S., Gowda, T.V. and Girish, K.S. 2006, 'The anti-snake venom properties of *Tamarindus indica* (leguminosae) seed extract', *Phytotherapy Research* (10), <http://dx.doi.org/10.1002/ptr.1951> accessed 3 June 2007.

Vanderbilt, T. 2005, 'When the Great White Way was the Hudson', *New York Times*, 29 May, <http://0-www.proquest.com.prospero.murdoch.edu.au:80/> accessed 5 September 2007.

Verghese, J. 1998, 'The snake that bit the hand that fed it', *Muscle & Nerve* (2), <http://dx.doi.org/10.1002/ (SICI)1097-4598(199802)21:2<259::AID-MUS18>3.0.CO;2-X> accessed 20 May 2007.

Vetter, R.S. and Schmidt, J.O. 2006, 'Semantics of toxinology', *Toxicon* (1), <http://www.sciencedirect.com/> accessed 8 June 2007.

Visser, L.E., Kyei-Faried, S. and Belcher, D.W. 2004, 'Protocol and monitoring to improve snake bite outcomes in rural Ghana', *Transactions of the Royal Society of Tropical Medicine and Hygiene* (5), <http://www.sciencedirect.com/> accessed 19 May 2007.

Warrell, D.A. 2006, 'Australian toxinology in a global context', *Toxicon* (7), <http://www.sciencedirect.com/> accessed 5 June 2007.

Watts, P.C., Buley, K.R., Sanderson, S., Boardman, W., Ciofi, C. and Gibson, R. 2006, 'Parthenogenesis in Komodo dragons', *Nature* (7122), <http://www.ncbi.nlm.nih.gov/entrez/query.fcgi?cmd=Retrieve&db=pubmed&dopt=Abstract&list_uids=17183308> accessed 21 August 2007.

White, J. 1998, 'Envenoming and antivenom use in Australia', *Toxicon* (11), <http://www.sciencedirect.com/> accessed 18 May 2007.

White, J. 2008, 'An unusual housemate', *Daily Mail Online*, 30 April 2008,

<http://moray.ml.duke.edu/projects/hippos/Newsletter/InTheNews.html#News3> accessed 2 May 2008.

Williams, D., Wuster, W. and Fry, B. G. 2006, 'The good, the bad and the ugly: Australian snake taxonomists and a history of the taxonomy of Australia's venomous snakes', *Toxicon* (7), <http://www.sciencedirect.com/> accessed 13 May 2007.

Williamson, J.A., Fenner, P.J., Burnett, J.W. and Rifkin, J.F. (eds) 1996, *Venomous and Poisonous Marine Animals: A medical and biological handbook*, Kensington, NSW, University of New South Wales Press.

Wuster, W., Dumbrell, A.J., Hay, C., Pook, C.E., Williams, D.J. and Fry, B.G. 2005, 'Snakes across the Strait: trans-Torresian phylogeographic relationships in three genera of Australasian snakes (Serpentes: Elapidae: *Acanthophis, Oxyuranus, and Pseudechis*)', *Molecular Phylogenetics and Evolution* (1), <http://www.sciencedirect.com/> accessed 1 May 2007.

—— Golay, P. and Warrell, D.A. 1997, 'Synopsis of recent developments in venomous snake systematics', *Toxicon* (3), <http://www.sciencedirect.com/> accessed 14 May 2007.

——1999, 'Synopsis of recent developments in venomous snake systematics, No. 3', *Toxicon* (8), <http://www.sciencedirect.com/> accessed 14 May 2007.

Young, B.A. and Kardong, K.V. 2007, 'Mechanisms controlling venom expulsion in the western diamondback rattlesnake, *Crotalus atrox*', *Journal of Experimental Zoology Part A: Ecological Genetics and Physiology* (1), <http://dx.doi.org/10.1002/jez.a.341> accessed 10 May 2007.

A NOTE ON MEASUREMENTS

For simplicity's sake, metric measurements have been used throughout *Deadly Beautiful*, except where original quoted material used imperial measurements. Some useful metric–imperial equivalents are listed below.

1 inch = 2.54 centimetres or 25.4 millimetres
1 foot = 30.4 centimetres
1 metre = 3.28 feet or 1.09 yards
1 kilometre = 0.621 miles
1 square kilometre = 0.386 square miles

1 ounce = 28.35 grams
1 pound = 454 grams
1 kilogram = 2.2 pounds
1 tonne (or metric ton) = 2200 pounds

1 litre = 0.264 gallons

Index

A

B

C

D

E

F

G

H

I

J

K

L

M

N

O

P

R

S

T

U

V

W

Y